Agency in Poverty and War

This book examines how people's self-awareness is affected by both internal and external factors amid war and poverty. It explores how agency has influenced the inward human development of rural women who face triple disadvantages related to gender, ethnicity, and access to economic power.

It presents a multidisciplinary perspective on the intersection of war and poverty through narratives of surviving women. It advances understandings of how rural people, peasants, and Indigenous peoples of Peru, particularly women, have experienced poverty and war as a combination of oppression, repression, and aggression. It explores how their agency is affected and how it evolves during and after conflict in their search for truth and justice. It does this by taking the capability approach combined with insights from perspectives on raising consciousness and inner transformation in human development in which awareness of rural people's experience enables them to be free and can move them from survival to conscious agents.

This book offers new narratives to evaluate the hazards of poverty and war and the potential human security for rural people agency and empowerment in building peace. It will be of interest to scholars and practitioners of development studies, peace and security, political Latin America geography, rural communities, peace and conflict studies, human development and political studies.

Vivianna Rodriguez Carreon, PhD, is a social scientist who has lectured in peace and conflict studies at The University of Sydney. She is a transdisciplinary scholar focusing on human development and exploring the inner and outer underpinning roots of human experience in human agency. As a trauma-informed action researcher, Vivianna is an international facilitator of awareness-based systems change.

Routledge Studies in Hazards, Disaster Risk and Climate Change

Series Editor: Ilan Kelman, Professor of Disasters and Health at the Institute for Risk and Disaster Reduction (IRDR) and the Institute for Global Health (IGH), University College London (UCL)

This series provides a forum for original and vibrant research. It offers contributions from each of these communities as well as innovative titles that examine the links between hazards, disasters and climate change, to bring these schools of thought closer together. This series promotes interdisciplinary scholarly work that is empirically and theoretically informed, with titles reflecting the wealth of research being undertaken in these diverse and exciting fields.

Climate Change and Risk in South and Southeast Asia
Sociopolitical Perspectives
Edited by Devendraraj Madhanagopal, Salim Momtaz

Gender-Based and Layered Disasters
Place, Culture and Survival
Nahid Rezwana & Rachel Pain

The Post-Earthquake City
Disaster and Recovery in Christchurch, New Zealand
Paul Cloke, David Conradson, Eric Pawson & Harvey C. Perkins

Local Adaptation to Climate Change in South India
Challenges and the Future in the Tsunami-hit Coastal Regions
Devendraraj Madhanagopal

Good Practices for Disaster Risk Management of Cultural Heritage
Practices of ITC participants
Edited by Rohit Jigyasu, Kim Dowon and Lata Shakya

Agency in Poverty and War
Consciousness in Rural Human Development
Vivianna Rodriguez Carreon

For more information about this series, please visit: https://www.routledge.com/Routledge-Studies-in-Hazards-Disaster-Risk-and-Climate-Change/book-series/HDC

Agency in Poverty and War

Consciousness in Rural Human Development

Vivianna Rodriguez Carreon

Routledge
Taylor & Francis Group
LONDON AND NEW YORK

First published 2025
by Routledge
4 Park Square, Milton Park, Abingdon, Oxon OX14 4RN

and by Routledge
605 Third Avenue, New York, NY 10158

Routledge is an imprint of the Taylor & Francis Group, an informa business

British Library Cataloguing-in-Publication Data
A catalogue record for this book is available from the British Library

Library of Congress Cataloging-in-Publication Data
Names: Carreon, Vivianna Rodriguez, author.
Title: Agency in poverty and war / Vivianna Rodriguez Carreon.
Description: Abingdon, Oxon ; New York, NY : Routledge, 2024. |
Series: Routledge studies in hazards, disasters and climate change adaptation | Includes bibliographical references and index.
Identifiers: LCCN 2024015493 (print) | LCCN 2024015494 (ebook) |
ISBN 9781032586526 (hardback) | ISBN 9781032586533 (paperback) |
ISBN 9781003450955 (ebook)
Subjects: LCSH: War and society. | Rural women--Social conditions
Classification: LCC HM554 .C39 2024 (print) | LCC HM554 (ebook) |
DDC 303.6/6--dc23/eng/20240430
LC record available at https://lccn.loc.gov/2024015493
LC ebook record available at https://lccn.loc.gov/2024015494

ISBN: 978-1-032-58652-6 (hbk)
ISBN: 978-1-032-58653-3 (pbk)
ISBN: 978-1-003-45095-5 (ebk)

DOI: 10.4324/9781003450955

Typeset in Times New Roman
by KnowledgeWorks Global Ltd.

To all the rural women living in sorrow of poverty and war, to contribute to their awakening of their *conscient agency*.

Contents

Figures

Preface

Writing this book restored my connection to my roots, the wisdom of oral culture, and its agency in the hologram of knowledge. It reflects the journey of identities and their positive and negative associations made in our consciousness that bring an assumed order to our intellect and yet fragment the wholeness of our humanity. I craft a new narrative to create a transdisciplinary discussion for scholars and practitioners working with people who have experienced the hazards of oppression in poverty, repression and aggression in war and are involved with wicked problems. Also, my analysis can benefit those taking part in projects of truth commissions and in periods of transitional justice mechanisms. The exploration opened in me, a development and peace action researcher, a new integrative way to see systems in the aftermath of armed conflict.

In getting ready for publishing, my courage in reviewing my writing didn't come without hesitation. When I opened these pages, I had concerns about whether I was loyal to my non-linear way of exploring this research. Yet, I see the gap in approaches to understanding the complexity of war urged incorporating, along with development, peace, and conflict studies, how victims-survivors carry their consciousness. A new paradigm metaphysically inspired by the invisible structures of perception enables us to see what is shaping the threats to safety in human agency. For example, I wondered if I could reframe the perception of the *campesinos*, the *peasants*. English is my second language, and when I was discussing it with my English advisor, I found out that a peasant identity does not have power in English literature. I wanted the peasant's identity to be perceived with a positive connotation. I only needed to investigate the Oxford dictionary: a peasant is a poor smallholder or agricultural labourer of low social status. I continued elaborating on the peasant's identity concept and asked myself, but what about if this is not how they identified themselves? My realisation that rural people already come to this world with identification through the lenses of the system of power domination constructing knowledge encouraged me to bring consciousness to my narrative. The identity associated with negative connotations fragments societies and reinforces the Othering. The identities' hierarchical order has compromised the sense of worthiness of their agency expression and identified them as "oppressed", but if the *campesinos'* dignity is seen within their idioms of oral culture flourishing, a new embodiment through empowerment ought to be seen.

Raising the consciousness for awareness is the opportunity to reconnect with vulnerability as the strength to support witnessing the experiences of oppression, repression, and aggression by gradually letting go of the need to be certain and inviting the freedom from the known. My research was to explore the empowerment process for identities perceived with triple disadvantage for their gender, ethnicity, and access to economic power in the midst of adversity, poverty, and war. Then, some doubts emerged in reviewing my chapters for publishing this book. Would my chapter about the repression in war be too raw, avoiding romanticising using the language of distance with more quantitative data and intellectual subtitles, and less the crude sensations internalised and working against agency? I can only hope this complex view of the inner and outer interconnections and how both dynamics interweave and pollinate across sectors are seen and experienced.

Not much has changed in the world to become more equal and peaceful since I finished my research. The statistics, if any, show our lack of capability to live in peace and without poverty. Humans cause the disasters of war, yet it seems beyond our scope to change. I remember when doing my research, I was tired of going to conferences where I heard over and over the statistics of what happened in war as if the "discoveries" were new. I witnessed the aftermath and investigated the literature through the usual lenses in the context of political violence, poverty, armed conflict, and transitional justice. I chose to focus on those living in the highlands and in the jungle. We shared the same landscape; I was born between both ecosystems. Hidden in the web of structural violence, underdevelopment, and insecurity, I contemplated that there was an internal human development process happening simultaneously for the participants who were building peace to survive. However, the faces of the victims-survivors in rural and isolated areas seemed to have an identikit. I tried to make sense of the inhuman dynamics seen in war and the multidimensional causes of oppression and effects of repression through the worldview of victims-survivors and how, in seeking peace, they experience a sense of injustice. I realised the literature was informative yet insufficient to make sense of people's unspeakable experiences. The implications of post-conflict and trauma tend to be treated separately and without the internalisation of the inner and outer phenomena. I saw how the war fragmented the peasant's identity and risked their connection with the land. The response to the gap in the traditional literature was consciousness.

It is hard to take the reader through a journey that gives justice to what happens in war. While the case study is about Peru, and most of the victims-survivors are women, the dynamics intertwined are shared among other rural areas experiencing conflict globally. If the hazards found in armed conflict are preventable, what is the real threat if we grow from survival thinking that counts the number of lives and deaths? In this book, we find that while survivors are alive in war, part of them dies. In going beyond traditional development and security, I invite you to read what I explored: the crucial identification of threats to human security through awareness of the disconnection in the capability to sense oneself, the *vital core,* which I called a hologram perception.

I am thankful for the number of mentors who helped me shape parts of the puzzle. In researching the post-war context of poverty in the mountains, I recaptured the teachings of the land where I grew up. The connections of the visceral experiences in-formed (inwardly formed) me with a meta capacity to consciously internalise knowledge. I mention here the scholars of the concepts who supported me, either through their mentorship or writing. In transitional justice, Dr Wendy Lambourne supported me when framing the period of the Truth and Reconciliation Commission (TRC) in Peru. In Paulo Freire's work on different levels of consciousness, I find out the inner qualities of periods in transition. I was inspired by the work of Professor Sabina Alkire on Amartya Sen's capability approach. In 2011, as a Visiting Student at the Oxford Poverty and Human Development Initiative (OPHI), through her mentorship, the identification of threats to human security became a central part of my research. It was then that I integrated my understandings of human development. The experiences of connection and disconnection with the inner self were beyond the literature of Western psychology. In my experience with the oral cultures, the Being was spiritual, which led me to immerse myself in the wisdom of the rural and isolated spaces and dive into Jiddu Krishnamurti's teachings. By all means, this work would not be possible without the interaction with the participants, the victims-survivors, and everyone named in the acknowledgement section who were part of this research expanded over ten years from 2004 to 2014, curated and published for you in 2024.

Acknowledgements

I want to acknowledge the victims-survivors whose stories are interwoven with the narrative of agency transforming inwardly. I want to thank my research supervisor, Dr Wendy Lambourne, who read this manuscript several times, for her endless dedication to mentoring, feedback, and unconditional support of the approaches used. I recognise my English advisor, Dr Bronwen Dyson, who, at the beginning of my work, supported me in finding the words I needed to convey to communicate the inner and outer dynamics of oral languages. I thank the supervisors I had in specific areas and times, in human rights Professor Danielle Celermajer at The University of Sydney during my doctorate candidature, and expert in human development and capabilities Professor Sabina Alkire at Oxford University during my visiting fellowship in Hilary and Trinity terms. The readers who assessed the scholarly value of this work are peacebuilding expert Professor Elisabeth Porter, post-conflict and transitional justice Advisor Félix Reátegui Carrillo, and Peruvian scholar Associate Professor Eliana Barrios Suarez. My former PhD colleagues who supported listening to the insights that sparked while sharing the office lounge and coffee moments at the Center for Peace and Conflict Studies were Leticia Anderson, Lydia Gitau, Lyn Dickens, David Penklis, James Tonny Dhizaala, and Eyal Mayroz. In particular, my co-sharing space friend Annie Herro and my former colleague friend, Punam Yadav, also without whom this book would not have been possible. Punam, Associate Professor at the University College London (UCL), knew the new human agency lenses I was bringing to approach armed conflicts and put me in contact with Routledge. I thank UCL Professor of Disasters and Health Ilan Kelman for reading my proposal and believing in this work. I am grateful for their editorial comments to author Adam Courtenay, and I thank Professors Emeritus Garry Trompf and Stuart Rees, and Associate Professor Salvatore Babones for reading and giving me feedback on my early work. I am also thankful for the number of members from organisations who helped me connect with the communities and centres of information in both rural and urban Peru. I also thank the number of non-government organisations, scholars, journalists, public servants, and practitioners whose knowledge was shared to enable my research. This list is not exhaustive of all the people who stepped up along the journey. Fellow colleagues and friends who supported me this last year when curating this publication and reading part of my edited work, from the

Timeless Wisdom Training (TWT) Lori Shridhare and Matthew Green, and from the Inner Development Goals (IDG) Fredrik Lindencrona. I also thank Routledge, Commissioner Faye Leerink, and all the publishing editors for their patience and support in believing this work is needed. Finally, I express my sincere gratitude to my husband, Shaun Rosenbaum, for taking care of our children, Maximiliano and Leonardo, during the many weekends and nights I spent late in the office in the early years of this research. I am truly thankful for their significant moral support leading up to the publication of this work.

List of Abbreviations and Acronyms

ANFASEP	Asociación Nacional de Familiares de Secuestrados, Detenidos y Desaparecidos del Perú (In English: National Association of Relatives of Kidnapped, Detained and Disappeared of Peru)
CAD	Comités de Autodefensa (In English: Self-Defense Committees)
CEDAW	Committee on the Elimination of Discrimination against Women
CMAN	La Comisión Multisectorial de Alto Nivel (In English: The Multisectorial Commission of Higher Level)
CONADEP	The National Commission on the Disappeared
CONAMUACAI	Coordinadora Nacional de Mujeres Afectadas por el Conflicto Armado Interno (In English: National Coordinator of Women Affected by the Internal Armed Conflict)
EPAF	Equipo Peruano de Antropología Forense (In English: Peruvian Forensic Anthropology Team)
FEDEFAM	Latin American Federation of Associations of Detained-Disappeared
INEI	National Institute of Statistics and Information
PCS	Consejeria de Proyectos (In English: Project Counselling Service)
PIR	Plan Integral de Reparaciones (In English: Comprehensive Reparations Plan)
UNDP	United Nations Development Programme

Part I

Introduction

The Inner and Outer Approach for Agency and Empowerment

1 The point of departure

Armed conflict in rural areas of Peru

A rural woman's story is the point of departure. Rosario was five years old in 1988, when late one afternoon, her father and uncles were taken by the military in the high jungle of Peru. They would never be seen again. Her mother, Maria Guimarita, walked every day for nearly an hour from her *caserío*, a rural hamlet, to the closest police station in San Martin de Alao to find out the whereabouts of her father. After a few failed attempts to get any information, Maria was threatened when she communicated about her intention to report the disappearance to the human rights organisation in Tarapoto.[1] Back then, the main town was two days by foot away from San Martin de Alao and roads could not be accessed by motorised transport. Poor roads are just one among many different wants of resources—being in lack of electricity and water was also experienced as a way of living in the *caserío*. Impelled to leave her three children under the care of her elderly parents, Maria Guimarita walked the two days on her way to find "justice", as Rosario described in her statement. This act of reporting the disappearances to a human rights organisation shows uncommon bravery among rural women, the kind of courage repeated over and over again by mothers during war.

Rosario's testimony tells how the night her mother returned to the *chacra* (the farm), her grandparents were tortured by "men with covered faces". The same night, Maria Guimarita, was taken away. After two days of searching, Rosario's mother was found by her elderly grandparents a hundred metres from the family's *chacra*. Her body showed signs of torture, and other grotesque marks of violence. Rosario's story was invisible to the public until a truth commission was formed to investigate the truth of those unnoticed during the conflict.

I ask myself here, unnoticed for whom?

According to the Peruvian Truth Commission's report (Comisión de Entrega de la Comisión de la Verdad y Reconciliación [CVR], 2004) the survivors most affected were the rural women living in economic poverty, recognising poverty as being mainly due to lack of access to water and electricity and limited financial resources (CVR, 2004). Women were also the most opposed to the conflict and mothers, like Maria Guimarita, were empowered by seeing injustice and acted despite being physically threatened. She had exercised her agency, what she valued and had a reason to value. The understanding of the term "agency" in the context

DOI: 10.4324/9781003450955-2

of war and poverty is integral to the meaning of empowerment. "The agency aspect is important in assessing what a person can do in line with his or her conception of the good" (Alkire, 2009, p. 487; Sen, 1985, p. 206). In Rosario's mother's case, the exertion of her agency comprised what she regarded important for her "conception of the good".

Searching for the stories of agency and empowerment

Rosario's file is one of more than a thousand recorded stories at the Centre of Information for Collective Memory and Human Rights (Centro de Información para la memoria colectiva y los derechos humanos), a documentation centre in Peru. This centre located in Lima was created to store and record the empirical data collected by the Truth and Reconciliation Commission (TRC) (Comisión de la Verdad y Reconciliación CVR). In visiting this centre, my purpose was to search for witness' reports from among the thousands of available public hearings transcripts and retrieve stories that strongly characterised the experiences of individual and collective agency in the midst of war and poverty. Rosario's story was the one that stood out for me. I was around the same age as her and lived in the same country at the same time, but here the similarities end. The Peru I experienced then and the one she had to endure were different realities. In the Commission's final report (CVR, 2004), her story represented one of the 17,000 individual accounts publicly recorded of the armed conflict.

While reviewing the TRC cases at the documentation centre, I began to understand that many women from remote areas, like Maria Guimarita, took an agentive role of motherhood that transcended during the conflict to a more significant participative role within their respective communities. Typified by the attempts to search for her husband by requesting aid from a human rights organisation, many women like her risked their own lives. Children who lived during this conflict are adults now, like Rosario herself, and they survived thanks to their mothers' protection and capacity for association. Rural women's active role, narrated in testimony after testimony, required the kind of instinctual agency and selflessness of thinking beyond their protection to participate in taking care of others.

Collecting data in rural areas: memories in the *chacra*

In reviewing the data about the Peruvian conflict, I found the lived experiences in rural areas were neglected. People living in poverty in the mountains of Ayacucho were the most affected. In Ayacucho's region more than 40% of the population of 69,000 died and disappeared between 1980 and 2000 (CVR, 2004, p. 21).[2] A national report on Peru showed that the rural population suffered the greatest poverty (INEI, 2012, p. 170) and that the levels of poverty in Peru are more acute for women than for men. For example, in 2009, the National Institute of Statistics and Information (Instituto Nacional de Estadística e Informática [INEI]) reported that 77.4% of all women lived in extreme poverty and they were from the rural Andean areas (INEI, 2009, p. 28).[3] During the 20 years of armed conflict, the vast number of those who died (85%) resided in Ayacucho or similar economically deprived regions. The United Nations Development Program (UNDP) pointed out that those states represented 9% of the income among the total families in Peru (CVR, 2004, p. 21). The violence of war within the country affected several regions in different ways. In rural Ayacucho, the highest intensity was registered in both numbers and depth.

In Ayacucho, those who died and/or disappeared shared similar identities in physical and cultural characteristics and experiences. According to the TRC, around 75% spoke Quechua and other native languages as their mother tongue, and almost 80% were men. Of this percentage, 66% were aged between 20 and 49 years old, and 75% were married or in a relationship (CVR, 2004, pp. 52–54). If we take into consideration the patriarchal structures which then prevailed in Peru's rural communities, we find that it was mainly rural men living in poverty—most with families and economic responsibilities to them—who died during the conflict. Therefore, rural Quechua women, widows and mothers were the greatest numbers of victims-survivors of the war, and their stories in-formed (shaped inwardly) the empirical and qualitative data.

What matters is creating new qualitative indicatives emerging from the quantitative data. The number of deaths is not just a number; it represents human experiences and changes in communities, and therefore knowledge which affects how the narratives of survivors' lives are affected individually and collectively. Who and what essentially were represented in the percentages? The argument I am making is constructed from the stories gleaned from information with different kind of quality. The common findings were that first-hand witnesses of the armed conflict had encyclopaedic memories, they were the carriers of oral languages, and they were rural women living in poverty.

I will make a pause here to highlight that I found in the literature texts referring to those with oral languages being known also as illiterate, a term which carries a negative connotation and synonymous with lack of education (Rodriguez Carreon & Vozniak, 2021). To change the narrative about Quechua and Aymara speakers, for example, I used Kelly's (2016) definition to address their knowledge as encyclopaedic memories. Oral cultures stored and encoded extensive amounts of information, conceiving memory in complex systems (more in Rodriguez Carreon & Vozniak, 2021).

Mamá Angélica's **story: behind empirical evidence of agency amid war**

The lived experiences emerging from the victim-survivors, like Rosario's story, drove my research to look beyond the memories amid war.

A well-known case is the story of late Mendoza de Ascarza in her search for truth and justice over her son's disappearance in Ayacucho. Her struggle is the journey of searching for answers for the whereabouts of her son. She became the *mamá Angélica* (mum Angelica), as people called her, for being the mother of many. From those disappearances and killings, many were left orphans and widows, and mothers without sons (Asociación Nacional de Familiares de Secuestrados,

Detenidos y Desaparecidos del Perú [ANFASEP], 2007). The empirical data in Angelica's story, her motivations for seeking truth and justice, shows the evidence of her agency amid poverty and war.

The empirical data from Rosario is from San Martin, a jungle forested area, and the story of Angelica is from a rural town of Ayacucho, situated in the highlands. Their stories are the starting point for understanding the empowerment process and agency in adversity. "Agency and empowerment are relational concepts" (Samman & Santos, 2009, p. 8). Empowerment is an intrinsic process conditioned by the individual's freedom to exert his or her agency (Ibrahim & Alkire, 2007). In that sense, agency, such as pursuing truth and justice in late Mendoza de Ascarza's case, is evolving, transforming and exposed if agents are free to act. I argue that agency is consciously exerted when awareness of the existing capabilities is free to notice them inwardly, enabling empowerment to flourish.

Locating empowerment within the process

Empowerment is a process. It is found in the interaction between the agents of change—the practitioner and the learner—by raising their common consciousness while entering a period of transformation, what Paulo Freire calls *concientização* (Freire, 1973, p. 134), usually translated as "conscientisation" in English. Empowerment happens through the raising of consciousness (White, 1994, p. 25). For instance, as Batliwala noticed, empowerment (in India) emerged first by shaping the meaning of the old intervention practice that treated women as a person to be assisted in creating "new spaces for women" to talk about their problems and "raise consciousness of their own sense of subordination". By this means, they could see by themselves the changes for which they aimed (Batliwala, 2007, para. 12.). However, actions to raise consciousness take time, and "it is a process that each individual has to do at her or his own pace" (Rowlands, 1995, p. 105).

This involves being *concientizado* (become conscious), the act of a person reaching a higher consciousness of their own experience. And for the purpose of my argument, I shall refer to this as conscientised, the act of being conscient. Consciousness in *conscient agency* is a process which cannot be reached by a person's thought patterns overnight. The interactive process of raising consciousness, and what it takes to be conscientised, relies on a thorough understanding of practice in the meaning of consciousness.

Consciousness is a concept defined robustly in the next chapter—Chapter 2: *Consciousness-capabilities* under the umbrella of human development and human security—as it is one of the pillars underpinning the process of formation and transformation happening within empowerment. The definition of consciousness used is multidisciplinary, a concept intersecting with, among others, awareness, wholeness, interconnection, and potentiality, from the educator Paulo Freire (1972, 1973, 1985, n.d.), the philosopher Jiddu Krishnamurti (1969, 1972, 1976, 1978, 1982, 1985), the neurophenomenologist Francisco Varela (1993), and physicist David

Bohm (1981, 1984, 1990, 1992, 2002). Consciousness is "to be conscious, not only of what is taking place around me but also to be conscious inwardly—what are my reactions, beliefs, fears, faiths, hopes, and all the forms of identification", and understand that there is "consciousness to all humanity, the content of which is the same" with just "physical and environment differences". For example, in relation to beliefs and suffering, people can believe in different things but everyone has beliefs; people can suffer for various reasons but all can relate to what is defined as suffering (Krishnamurti, 1985, pp. 142–148). Empowerment in this sense can emerge from the common understanding of the agent's awareness and inwardly of the common consciousness. The exploration needed is on what quality of space the system needs to enable empowerment to impact a person's consciousness living amid the hazards of war and poverty.

I question whether the flourishing of the empowerment formation is held back because the roots of a conscious paradigm are conditioned by the system. Whether, as Krishnamurti (1985) observed, we depend on our nationality, institutions, or relationships to such a degree that agents cannot raise consciousness, as their freedom to exercise their agency's capabilities to do this is constrained. If that is valid, do capabilities then depend on understanding the core of the freedom of agency? In Chapter 2, I will deconstruct the concept of capabilities in detail because, together with consciousness, they comprise the pillars for understanding the *conscient agency*, which I see as a necessary action for empowerment formation.

The relationship between empowerment and agency

Approaches to human agency focusing on human development (HD) and human security (HS) are based on the concept of capabilities. Both theoretical perspectives provide the umbrella for perceiving the threats of poverty and war affecting rural people's freedom of agency by looking through the lenses of HD and HS, respectively. In this book, I delved into Amartya Sen's pioneering work on the capability approach (1985, 1995, 1999, 2000, 2007, 2008, 2010), and other scholars who expanded the capabilities and human development definition, including Martha Nussbaum (2000, 2011), Sabina Alkire (2002, 2003, 2009, 2010) and Severine Deneulin (2006, 2009, 2011). For Sen, capabilities is enabling opportunities means to do things or to be someone with freedom (Samman & Santos, 2009, p. 4). He argues that a "freedom-centered perspective" explains how people pursue their goals and that the ends and means are viewed through the process of capabilities (Sen, 1999, p. 53). "The people have to be seen, in this perspective ['the perspective of freedom at the center of the stage'], as being actively involved—given the opportunity—in shaping their own destiny, and not just as passive recipients" creating their own human capabilities (Alkire & Deneulin, 2009, p. 27; Sen, 1999, p. 53). Furthermore, empowerment is a process from the interaction of both "agency" and "opportunity structure", in other words between the "ability" to make choices determined by the capabilities and the institutions of "social and political structures" (Samman & Santos, 2009, p. 3). Therefore, institutions can influence people's consciousness and condition their freedom of agency, for which capabilities are necessary for raising awareness to enhance empowerment.

Building and finding the argument: the bottleneck of agency

I considered the understanding of empowerment as the experiences of the rural women coming through their participatory agency, named here *active agency*. Although they were not recognised for their participation in decision-making by the system, it is noticed that rural women's passive agency changed inwardly during conflict into a more active role publically.

There was a "bottleneck", this term came from an interview in 2009 with a government official, Isabel Coral Cordero. The bottleneck prevented the agency process of rural women's empowerment to flow. Therefore, to be aware of the obstacles which oversee what held back women's agency for decision-making, I analysed the main points interacting between agency and the system, which led me to build the concepts, as well as the definition of the problem. The acupuncture points linked were agency, participation, and decision-making in the context of seeking the truth to get justice while living in poverty and war.

To find out what is behind the bottleneck, I sought the different observations made of victims-survivors by a number of scholars who have conducted fieldwork through the scope of their different disciplines: Kimberly Theidon (2004); Duncan Pedersen, Hanna Kienzer, and Jeffrey Gamarra (2010); and Eliana Suarez (2011) to mention a few. They all researched in Ayacucho. I highlighted their research on rural people's memory and representation of the meaning of violence through Quechua's expression of distress. It is in their experience of violence that their beliefs and thoughts embodied their memory and consequently their perceptions of life. I also analysed empirical data reflected in the changes faced by the people of Ayacucho, the work of Adam Webb (2011), and Gonzalo Gamio's (2009) research linking memory and justice. The conflict as a matter of justice in the Truth and Reconciliation Commission has been analysed by different scholars, including Lisa Laplante (2007), Julie Guillerot (2006, 2007), and Julissa Mantilla (2002, 2013); and the "veteran" scholars of Peru before, during, and after conflict, Carlos Ivan Degregori (1998, 1999, 2003), Nelson Manrique (1998), and Ponciano del Pino (1998, 2003). My fieldwork is supported by in-depth interviews and valuable insights from Narda Henriquez Ayin and Isabel Coral Cordero. They both researched people from Ayacucho during the conflict. A number of non-government organisation (NGO), practitioners, and other scholars had also researched the Peruvian conflict from different perspectives.[4] The aim in the integration of different disciplines is to attempt to see the concept of agency in poverty and war as a whole rather than as a fragmented lived experience to offer an integrative approach to development, peace, and security. In doing this transdisciplinary work, I acknowledged the inability to go into detail in areas that may deserve further attention. The purpose is to identify the threats to the freedom of human agency's *vital core*.

Under the scope focused on the human being as the centre for freedom, my argument is that:

> Freedom of the *vital core* through the consciousness and capabilities perceptions enhances the experiences of rural people's empowerment while transforming their oral world, by the acknowledgement of their agency in participating when seeking truth and justice, and by the recognition of the barriers for decision-making, in context, living in poverty and war.

Approaches to identify the threat to freedom

HD and HS provide the umbrella to perceive the threats affecting women's freedom in their agency. "Human development puts people back at the centre stage" (Streeten, 1995, p. IX). Human security recognises the "core aspects", the "capabilities or freedoms" of people (Alkire, 2010, p. 36) to perceive people free from poverty (freedom from want), and free from conflict (freedom from fear) (Tadjbakhsh & Chenoy, 2009). Physical and environmental differences have implications (Krishnamurti, 1985) that threaten freedom from doing things or being someone regarded as important. The impact on the human experience of being exposed to an environment of fear is analysed from the manifestation of the everyday oppression in poverty and repression in war. The contextual implications in peoples' lives provide the understanding of hidden threats. I found there is not much focus in the literature on contextual implications for people's agency potentiality. Yet, individual indicators (social, economic, political, and personal) are more influential in gender, for example (Rosen, 2007, p. 1). The context of poverty threatens the dignity of people, and finding and addressing that threat is central to HS (Tadjbakhsh & Chenoy, 2009, p. 9). The differences in gender inequality during conflict and in the aftermath brings further divisions based on race, class, gender, religion, and more (El Jack, 2003, p. 8).

How can the threats to security be identified and addressed to prevent deepening hazards? Security can be defined as obtaining freedoms, being free from danger, "feeling" safe (ODE, 2003; Tadjbakhsh & Chenoy, 2009, p. 10), and the feeling of freedom can be related to the experience of human dignity. But dignity is innate, no-one can take your dignity, and dignity has been defined as "the state of worthiness" (ODE, 2003), so all humans are worthy beings. However, I argue that freedom to feel worth depends on a number of circumstances. The exposure to what it means to be living in poverty and war can repress humans to the point of doubting not just their feeling of worthiness, as the TRC reported, but their feeling of being humans (CVR, 2004). Rural women emerging from violent conflict after experiencing war normally have in common a historical context of "sourness" (Porter, 2007, p. 11). This sourness is rooted so deeply that it can threaten vital feelings that are at the core of a human being with freedom. The identification of threats to this freedom is proposed to be found through the human security approach. As Sabina Alkire (2003) stated "human security takes its shape from the human being: the vital core that is to be secure" (p. 24).

In that sense, this research project explored the real threat to the freedom of human agency. The problem is when the *vital core* is deprived of its existential agency, which consequently hinders the empowerment formation. Through consciousness and capabilities perception, freedom of the *vital core* augments rural people's empowerment experiences while transforming their oral world when the system acknowledges their agency in participating and recognises the barriers to decision-making in their context.

I argue that empowerment is an inner dynamic forming agency in which the *vital core*, or sense of inner self, is in relationship from within the person and from without—the influences that bear on it from the outside world. Human agency connection with the *vital core* is reinforced by acknowledgement or insidiously suppressed through oppression and repression. The agency is always present in the empowerment process. When the connection with its *vital core* is heavily compromised, then agency enters a period of stagnation in its consciousness for connection. Their capabilities no longer have space for passive or active transformation but to survive. Empowerment continues forming and transforming as long as the vital core is not affected and disconnected from its sense of agency.

I describe four conceptualisations of agency for the analysis to support the understanding when living in poverty and war:

Passive agency: The accumulation of core beliefs in a person's life experience. Thoughts and capacities become embedded. Human agency is adapting its vital core.

Survival agency: Disconnection of core beliefs within the experience. Thoughts and capacities become compromised. Human agency is abjecting fragmentation of its vital core.

Active agency: Exertion of core beliefs. Thoughts and capacities become embodied. Human agency is resilient in its vital core.

Conscient agency: Awareness of core beliefs within the experience. Thoughts and capacities become acknowledged. Human agency is transforming its vital core.

The problem in empowerment

People's participation in *conscient agency* is the awareness of being part of the empowerment process by taking decisions about their own destiny. Participation alone by women at the grassroots level is essential for their empowerment and a necessary step to take for them to be involved in development policies (Karl, 1995, p. 110). However, White (1994) cautions against using participation as a key term because it is "kaleidoscopic" and changes are depending on the perception of the "eye of the beholder" and by the "hand of the power holder". Therefore, in places where the control of the project is in the hands of practitioners and/or those perceived with power over, it is necessary to distinguish two types of participation: pseudo participation and genuine participation. The first one gives the sense of "being present to listen to what would have be done *unto* them"; the second one gives the sense of being in partnership, deciding and empowered

(p. 17). The role of genuine participation described by White is the kind of participation necessary for empowerment to evolve. It resembles a participation based on awareness which relates to the concept of *conscient agency*. Empowerment is about the agent participating in decision-making which affects "her or his action and the condition of that action" (Young, 2011, p. 251).

People's decision-making when living in poverty within post-conflict scenarios, however, remains a utopia. The United Nations Security Council Resolution 1325 (2000) explains that the "role of women" is to participate and increase their decision-making in preventing and resolving conflicts (Suthanthiraraj & Ayo, 2010, p. IX). For post-conflict scenarios, obstacles to women's participation in decision-making are due to poverty, low literacy, and cultural issues.[5] Historically, women have been absent in government structures and decision-making processes because they are limited to occupying roles mainly in the private and domestic sphere. The execution of structured public policies has largely been a male-gendered occurrence. In the aftermath of war, women have to find their own way to fit into a male-dominated system (Giercycz, 2001, p. 23).

Government's recognition and acknowledgement of empowerment from women's participation in decision-making impacting their families and their communities does not happen. Participation across institutions is affected by a decision-making structure that excludes rural women. Sen (1999) identified the role of empowerment in women's lives: "the acknowledgement that women's power … can have far-reaching impacts on the forces and organizing principles that govern divisions *within* the family and society as a whole" (p. 192). Under that scope, empowerment is perceived when recognition and acknowledgement empower women to act and to challenge "the fundamental imbalances of power in their lives in the domains of the family and the community" (Subramaniam, 2006, p. 7). Porter (2007) extends this argument on the importance of acknowledging and being open to differences. The focus on war disrupts the recognition of the need to embrace the diversity that women can offer for peacebuilding. But when conflict deepens, the root cause of violence of identity clashes exacerbates the differences. Then, if women are not aware of their "intrinsic capabilities", they tend not to "seek recognition of their particularity or specific needs or try to be included in decision-making bodies" (Porter, 2007, p. 69). Hence my focus is to identify the threats to women's awareness of their intrinsic capabilities. What is it that women need to be free from in order for them to be able to identify internally the opportunity *to do* things or *to be* someone for participation and decision-making?

Contextualising the problem: beyond case studies

People living in war zones are often highly traumatised at physical, cognitive, and emotional levels, and their struggles for survival restructure the inner and outer of their agency roles. Trauma is a fragmentation (Hübl, 2020), a broken experience (Van der Kolk, 2014; Herman, 1997), which expression disconnects the self from the self (Rodriguez Carreon & Vozniak, 2021, p. 36). According to Herman (1997), its impact is too terrible to "banish them from consciousness" to the point of being unspeakable (p. 1). The extensive impact of armed conflict leaves a large number of widows and mothers of those who disappeared. They suffer the long-term implications of not knowing the whereabouts of their husbands or sons by "not being able to bury their loved ones and mourn properly" and because they lack recognition of their "widowhood" status (Lindsey, 2001). The lack of recognition of their status as widows affects them in different dimensions like their rights to land, inheritance, remarriage, among many other losses.

Women in conflict were witnesses to their tragedy; they had to change their roles and were economically and culturally affected in their families and their communities overnight, a pattern that is found in many conflicts—mothers searching for their children or grieving widows from war zones. This is not unique to Peru's armed conflict. From the Nepal war, the organisation Women for Human Rights found 67% of the widows were between 20 and 30 years of age (Thapa, 2007). In July 1995, during the Srebrenica occupation (in the former Yugoslavia), between 7,000 and 8,000 men were executed. The Mothers of Srebrenica—who had very little education and no access to power and decision-making bodies—by associating themselves, became an example of how women can reinvent their role in a post-conflict society to become agents of change (Simic, 2009). Women face dramatic challenges in their roles, becoming, as Sen (1999) describes, active agents of change (p. 189). At the same time, rural women's intrinsic characteristics of human agency involve coping with the consequences of their war exposure, which ruptures the movement of their agency development into a conscious stage.

Coping with and surviving hardship are agentive roles because it is within this intrinsic process of struggle that change occurs, *passively* in the outer and *actively* in the inner self. The women living in poverty coped with the underdeveloped environment and during war become targets of sexual abuse, threatening their agency and ability to cope. "Women are raped in all forms in armed conflict, international or internal, whether the conflict is fought primarily on religious, ethnic, political or nationalist grounds, or a combination of all these. They are raped by men from all sides – both enemy and 'friendly' forces" (Chinkin, 1994). There is enough evidence from an Amnesty International report of several contexts negatively affecting women, from surviving the consequences of rape and sexual violence to the deaths and disappearances of families and friends. Also, the report describes how the impunity allowed to the perpetrators is a difficult experience to cope with for victims and their families.[6] Confronted living with rape and disappearance, rural women's capacity to cope submit to *surviving agency*.

In Peru, the triple hazards exposure to structural, direct, and ongoing violence of poverty, war, and seeking the unknown for justice threatened the *vital core* of rural women, challenging their freedom by submerging their agency to survival participation. It is found that women were threatened, raped, and murdered by both the military and the Shining Path (Chinkin, 1994). Living in a life-threatening environment, women underwent an "intensive learning process", acquiring and consolidating different skills (Guillerot, 2006, p. 143) of survival. Nevertheless, for many women in Ayacucho, poverty and conflict had severe implications and manifested "expressions of distress" such the *llaki*, a Quechua word for "sorrow, sadness" (Pedersen, Kienzer & Gamarra, 2010), and craziness (Theidon, 2004). In the highlands, the leaders presented problems, the consequences of their constant sensibility and exhaustion, which troubled their minds (Coral Cordero, 1996, p. 54; 1998). After exposure to hardship in some cases, women presented resilience and endurance in the face of their trauma and Quechua women were seen participating in association to help their communities (Suarez, 2011). I challenge the meaning of resilience in the literature, where endurance is seen from the outside as a capacity to function. People's mind might not be compromised, yet inwardly, their body might be emotionless, and the person continues participating and doing without being. Without knowing what is happening with the inner Self, a person's vulnerability is "hijacked", and therefore, their participation is disconnected from their consciousness.

Among the participants' stories in this chapter, Rosario represented a woman who endured survival and impunity for her mother's torture and murder, as a consequential implication of living in poverty and conflict. On the other hand, Angelica represented the agent, who in the search for justice, drastically changed her roles from domestic and private to public. While the stage of being "lost in thoughts" at times was acknowledged by both during my interviews, this is not recognised as a constant mind stage. For that reason, I argue that their *vital core* was not in a complete sense compromised, as those with *llaki*. Angelica's story of search became the driving motivation of her agency to participate and, together with other women, her experiences were embodied during times of hardship. This research embeds their experiences to inform the inquiry on the threats to the development of their agency's inner condition.

The method of unpacking the problem

Women's survival stories like the late Mendoza de Ascarza's case are the driving motivation to find out the key indicators for human agency amid adversity as well as finding out the root of the problem. How can we conceive that ignoring the effects on the inner self is a problem, and that it comes from failing to acknowledge the essential contributions of rural people's oral culture to the process of empowerment in agency development? It was necessary to do the unpacking approach found in Paulo Freire's (1985) work. Breaking down the theme into concepts, looking deeply into the meaning of each concept and at the same time maintaining context. The process of unpacking from the initial stage of the story was essential. Freire said:

> We admire the words of our proposed theme so that in looking at them from the inside we can recognize that they ought not to be dealt with as a mere cliché. This theme is not a slogan; it's a problem in itself and a challenge.
>
> While just looking at these words and so remaining in their periphery, we might limit ourselves to talking about their message as one of "ready made ideas".
>
> In critically and inwardly analysing this message we have a chance to break down its constituent parts. Dividing the totality into its components enables us to return to the totality with a more in-depth comprehension of its meaning.
>
> (Freire, 1985, p. 38)

Applying Freire's statement, each word was used to formulate a whole thrust. Each word carries its own individual and legitimate meaning; it contains a whole concept in which definitions carry the solution and similarly its own problem. Each word carries an extended meaning once linked to other concepts within the main topic and afterwards, when placed in different contexts. For example, if I evaluate *agency* solely in *armed conflict*, it will open a different thrust than if I link the words *agency* and *participation* in *poverty*. The research behind this book acknowledges the interdependence of the context for those living in poverty and war and their connection with seeking truth and justice. At the same time, acknowledging that once experiences are conceptualised creates distance between the experience and the concept. Therefore, the analysis of each individual concept recognising their interdependence's lived experiences within context is necessary. While reading about development, I encountered a problem with its conceptualisation. The development's contextualisation depended on the cultural eye of the beholder. Also, I found the conceptualisation of being a woman blended in several contexts of poverty and war; however, women's experiences were sorted and compartmentalised. For example, the concept of being a woman is either found as just woman in development or just woman in conflict. The challenge here was to respect the interdependence of context and consider that each word or concept carries individual problems within the framework of a bigger problem or question.

Combining each concept and one dimension helped to conceptualise the structure of victims-survivors' lived experiences and to notice the interdependence between the environment or system as a whole in relationship to the singularity of human experience, their living day-to-day. The inner and outer qualities emerging from the information given in the stories described the problem.

The qualitative data from the stories will "bring to life" each concept's meaning in each problem, as Paulo Freire remarks. For example, Chapter 4 "War and its rawness through cruelty and repression" about the armed conflict, describe the causes and consequences of repression, underdevelopment, and injustice for people living in poverty and, at the same time, what they encountered living in war. The methodology of unpacking includes accounting for the stories collected to represent the problem in each chapter. The analysis of the empirical findings under the umbrella of the different approaches, consciousness and capabilities, helps to illuminate a new perception.

Steps taken to self-awareness

As a Peruvian, awareness of my life experiences living in my early years in rural areas contributed to my connection with local knowledge and make it possible to concentrate with what was happening on the ground. In the field, the problem is enriched and augmented by the experience (Madison, 2005, p. 21). I had the opportunity to live in different environments in Peru: the jungle of the Andes mountains and the urban city of Lima. I had inner knowledge from ways of being to empathise and understand deeper participants' lived experiences in provinces and cities. My experience in-formed me with a unique perspective to find geographic similarities and local ways of living of victims-survivors. This self-reflection on the knowledge of acting in different situations and environments gave me action confidence. Action confidence is an embodied experience with significant epistemological change in one's perception for the purpose of action (Pomeroy & Oliver, 2021). This transformative sense of learning involved to become familiar with the research environment and at a deeper level, putting empathy and listening into practice.

Before connecting with participants through dialogue, one needs to consider one's own self-awareness as a potential medium for knowledge transformation. An evaluation of self-awareness is thought to provide accountability for our own moral responsibility, relative to representation and interpretation (Madison, 2005, p. 7). A central aspect of interacting with participants is the ability to trust. I listened deeper to their stories and observed their daily living events while situating[7] myself in their space. This practice helped me to empathise, thereby offering my trust and at the same time, be able to honour their trust.

Different traditional methods for qualitative research included grounded theory, critical and post-critical ethnography when collecting data, or as I prefer to say, when connecting with the people and their qualities that created the content. During the trips to the field either the highlands or the jungle, I applied critical ethnography, respecting the subjects and acknowledging they possess agency and history. An important question for a critical ethnographer to address before getting into the field is who is the researcher? (Madison, 2005, p. 23).

> Critical ethnography begins with an ethical responsibility, [means] a compelling sense of duty and commitment based on moral principles of human freedom and well-being, and hence a compassion for the suffering of living beings … as a result, the researcher feels a moral obligation to make contribution toward changing those conditions greater toward freedom and equity … this means the critical ethnographer contributes to emancipatory knowledge and discourses of social justice.
>
> (Madison, 2005, p. 5)

There is a shift when framing our position as researchers in acknowledging fieldwork as personal experience, where our "intuitions, sense and emotions" are effectively blended into and become an inseparable part of the transformation (Madison, 2005, p. 8). In acknowledging there is an effect on the participants and the researcher, I became aware of possible changes that might occur from dialogues with the Self, the Other, and the Environment. A dialogue is an action that could become generative for participants and researchers to create different ideas and meanings. Through these encounters of reflection, the inquiry continued to be constantly transformed, reviewed, or reassured.

When applying dialogue and self-awareness, I noticed that fieldwork in the aftermath of war involved interactions in which human beings shared their private information. This type of encounters between participants and researcher reveals a new quality of information, a breath of trust. My approach when listening to participants on several points during interviews imparts qualities of being where I am, in the here and the now, to inform a new perception. Awareness became a reflection in action, as Madison states (2005), an "engagement between human beings" (p. 14).

Each story is unique and with a possibility for a new meaning of agency to expand. My acquaintance with conflict experiences arose after reading stories like Rosario. Then with *mamá Angélica*, a notable case of agency to many, was the opportunity to narrate how one rural woman made a difference. The dimensionality of the interviewees' stories is encapsulated in the narrative that informs new connections and transforms its meaning. While, I noticed a story, I see isolated happenings. These incidents were coded to conform a new concept. For Glaser and Strauss (1999) the researcher's own experiences, gained through reading, watching, or those gained from other stories, bring richness to the creation of concept's categories (p. 67). For example, my lived experience in the field raised my awareness of the lack of security in transport and communications in rural areas and informed my contextual knowledge of what is to live in isolated and underdeveloped remote communities. My experience as a researcher was in post-war within the environment of underdevelopment and not living in poverty, not the same as when I was living in conflict as a non-researcher interacting with development; when listening to the experiences of rural women, it contained this multi-dimensionality of contextual hologram.

$$\text{Experiencing}\left(\text{inner and outer ways of being and doing}\right)$$
$$= \text{armed conflict} + \text{underdevelopment} + \text{living in poverty}.$$

According to the TRC report, the war left many wounds in victims-survivors. The inner and outer qualities of their experiences became the root and foundation for this research. Rural women overcame existing difficulties from the triple disadvantage identities connoted as powerless for their ethnicity, their context of living in poverty (class), and their gender. To clarify, those categories from their identities own not to be connoted negatively or powerless. The "disadvantage" shows the power dynamics existing during research and how the empowerment process manifested for participation in decision-making at the informal level. Madison argues that data carries the necessity to apply "ethical responsibility" and a "sense of duty". A move from "what is" to "what could be" (2005, p. 5) is interpreted as what is agency and what could be agency. The sense of duty is my ethical responsibility to support agency transformation in rural living.

Concluding with practical aspects of establishing connections with the participants and the dialogue itself, the way of being includes adventurous and mindful curiosity inner qualities of agency. The actions taken before and during fieldwork involved common sense and intuition to help measure the possibilities and probabilities of getting information. For logistics, for example, there was a necessary reliance on using local transport and needing to search for local informants to connect with participants out in the field. Being open to restructuring my perception of enquiries while in the field was a necessary part of my approach to collecting data as an action researcher. I also considered that all logistical impasses carried a meaning. To sum up, this research constantly saw the problem within context—space and time interaction intertwined with the past. All meanings contributed to part of the problem and part of the answer and were equally a part of the knowledge needed for the chosen method.

Notes

1 Tarapoto is located in the high jungle of Peru.
2 The data is based on the Truth and Reconciliation Commission's report. *Hatun Willakuy*, in the abbreviated version of the final report, points to an estimation based on the testimonies: around 23,969 Peruvians were reported to have died or disappeared during the conflict. Considering the number of people who did not give testimony and most did not have an ID, the number is 69,000 (CVR, 2004).
3 See more INEI (2000), which discusses the different meanings of poverty levels in Peru according to the capacity to access the price of basic food, goods and services.
4 The NGOs include: *Asociación Pro Derechos Humanos* (APRODEH; Peruvian Association for Human Rights), *Equipo Peruano de Antropología Forense* (EPAF; Peruvian Forensic Anthropology Team), *Para Que No Se Repita* (PQNSR; Never Again), *Consejería en Proyectos* (PCS; Project Counselling Service), and *Coordinadora Nacional de Mujeres Afectadas por la Violencia Política* (CONAMUACAI; National Coordination Group for Women Affected by Political Violence).
5 Office of the Special Adviser on Gender Issues and Advancement of Women, Department of Political Affairs (2004), *Enhancing women's participation in electoral processes in post-conflict countries.* EGM/ELEC/2004/REPORT.
6 Amnesty International (AI), *Women in Post-Conflict Situations: A Fact Sheet.* http://www.amnestyusa.org/violence-against-women/stop-violence-against-women-svaw/violence-in-post-conflict-situations/page.do?id=1108238
7 See more, *Situated Knowledges* from Donna Haraway (1988).

References

Alkire, S. (2002, February 16). *Conceptual framework for human security* (Excerpt: Working definition and executive summary). http://www.unocha.org/humansecurity/chs/activities/outreach/frame.pdf

Alkire, S. (2003). *A conceptual framework for human security*. In *Centre for research in inequality, human security and ethnicity (CRISE)*. University of Oxford.

Alkire, S. (2009). Amartya Sen. In P. Jan & S. Irene van (Eds.), *Handbook of economics and ethics* (pp. 484–492). Edward Elgar.

Alkire, S. (2010, May 1). *Human development: Definitions, critiques, and related concepts* (OPHI Working Paper No. 36). http://ssrn.com/abstract=1815263 or http://dx.doi.org/10.2139/ssrn.1815263

Alkire, S., & Deneulin, S. (2009). The human development and capability approach. In S. Deneulin & L. Shahani (Eds.), *An introduction to the human development and capability approach: Freedom and agency* (pp. 22–48). Earthscan.

ANFASEP Asociación Nacional de Familiares de Secuestrados, Detenidos y Desaparecidos del Perú (2007). *¿Hasta cuándo tu silencio? Testimonios de dolor y coraje* [How long will your silence be? Testimonies of pain and courage]. Ayacucho. ISBN: 978-603-45123-0-6

Batliwala, S. (2007). *Putting power back into empowerment.* 50.50 Inclusive Democracy. http://www.opendemocracy.net/article/putting_power_back_into_empowerment_0

Bohm, D. (1981). Insight, Knowledge, Science, and Human Values. *Teachers College Record*, 82(3), 1–16. https://doi.org/10.1177/016146818108200305
Bohm, D. (1984) In *John Cobb and David Bohm Interview, 1984* [Video]. Center for Process Studies (2020, November 20). https://www.youtube.com/watch?v=l7VKGcmWVCw
Bohm, D. (1990). A new theory of the relationship of mind and matter. *Philosophical Psychology*, 3:2-3, 271-286, https://doi.org/10.1080/09515089008573004
Bohm, D. (1992). *Thought as a system*. David Bohm Seminars.
Bohm, D. (2002). *Wholeness and the implicate order*. Routledge.
Chinkin, C. (1994). *Rape and sexual abuse of women in international law*. Symposium: The Yugoslav Crisis: New International Law Issues. Issue Vol. 5, No. 1. http://www.ejil.org/pdfs/5/1/1246.pdf
Coral Cordero, I. (1996). *Las mujeres en la guerra: impacto y respuestas* [Women in war: impact and responses]. CEPRODEP. Centro de promoción y desarrollo poblacional.
Coral Cordero, I. (1998). Women in war: Impact and responses. In S. J. Stern (Ed.), *Shining and other paths: War and society in Peru, 1980–1995*. Duke University Press.
Comisión de Entrega de la Comisión de la Verdad y Reconciliación. (2004). *Hatun willakuy* [The great story]: Versión abreviada del Informe Final de la Comisión de la Verdad y Reconciliación [Abbreviated Version of the Final Report of the Truth and Reconciliation Commission]. Navarrete.
Degregori, C. I. (1998). Harvesting storms: Peasant rondas and the defeat of Sendero Luminoso in Ayacucho. In S. J. Stern (Ed.), *Shining and other paths: War and society in Peru, 1980–1995*. Duke University Press.
Degregori, C. I. (1999). Reflections. In C. Arnson (Ed.), *Comparative peace processes in Latin America*. Stanford University Press.
Degregori, C. I. (2003). Introducción. In C. I. Degregori (Ed.), *Memorias de la represión: Jamás tan cerca arremetió lo lejos*. [Memories of repression: Never the so close attacked so far], (pp. 15–26). IEP.
Del Pino, P. (1998). Family, culture and "revolution": Everyday life with Sendero Luminoso. In S. J. Stern (Ed.), *Shining and other paths: War and society in Peru, 1980–1995*. Duke University Press.
Del Pino, P. (2003). Uchuraccay: Memoria y prepresentacion de la violencia política en los Andes. In C. I. Degregori (Ed.), *Memorias de la represión: Jamás tan cerca arremetió lo lejos*. [Memories of repression: Never the so close attacked so far], (pp. 49–93). IEP.
Deneulin, S. (2006). *The capability approach and the praxis of development*. Palgrave Macmillan.
Deneulin, S. (2009). Ideas related to human development. In S. Deneulin & L. Shahanied (Eds.), *An introduction to the human development and capability approach: Freedom and agency*. Human Development and Capability Association. Earthscan.
Deneulin, S. (2011). Advancing human development: Values, groups, power, conflict. In V. Fitzgerald, J. Heyer, & R. Thorp (Eds.), *Overcoming the persistence of inequality and poverty* (pp. 127–148). Palgrave Macmillan. http://opus.bath.ac.uk/15953/
El Jack, A. (2003). *Gender and armed conflict: Overview report*. In *Institute of development studies. BRIDGE team*. University of Sussex.
Freire, P. (n.d.). *Cultural action for freedom*. http://www.thinkingtogether.org/rcream/archive/110/CulturalAction.pdf
Freire, P. (1972). *The pedagogy of the oppressed* (B. Ramos, Trans.). Penguin Books.
Freire, P. (1973). *Education for critical consciousness*. Sheed and Ward.
Freire, P. (1985). *The politics of education: Culture, power and liberation*. Bergin & Garven.
Gamio Gehri, G. (2009). *Tiempo de memoria: Reflexiones sobre derechos humanos y justicia transicional* [The time of memory: reflections on human rights and transitional justice]. Centro de Estudios y Publicaciones.

Giercycz, D. (2001). Women, peace and the United nations: Beyond Beijing. In S. Inger & S. Dan (Eds.), *Gender, peace and conflict*. Peace Research Institute Oslo.

Glaser, B., & Strauss, A. (1999). *Discovery of Grounded Theory: Strategies for Qualitative Research* (1st ed.). Routledge. https://doi.org/10.4324/9780203793206

Guillerot, J. (2006). Linking gender and reparations in Peru: A failed opportunity. In R. Rubio-Marin (Ed.), *What happened to the women? Gender and reparations for human rights violations*. International Center for Transitional Justice, Advancing Transitional Justice Series. Social Science Research Council.

Guillerot, J. (2007). *Para no olvidarlas más: Mujeres y reparaciones en el Perú* [To do not forget them anymore: Women and reparations in Peru]. APRODEH, DEMUS, PCS.

Haraway, D. (1988). Situated Knowledges: The Science Question. In *Feminism and the Privilege of Partial Perspective. Feminist Studies*, 14(3), 575–599. https://doi.org/10.2307/3178066

Herman, J. L. (1997). *Trauma and recovery: The aftermath of violence – from domestic abuse to political terror*. Basic Books.

Hübl, T. (2020). *Healing collective trauma: A process for integrating our intergenerational & cultural wounds*. Sounds True.

Ibrahim, S., & Alkire, S. (2007). Agency and empowerment: A proposal for international comparable indicators. In *Oxford development studies* (Vol. 35, No. 4, pp. 379–404). Routledge. http://www.tandfonline.com/doi/pdf/10.1080/13600810701701897

Instituto Nacional de Estadística e Informática. (2000, January). Metodología para la Medición de la Pobreza en el Perú [Methodology for Measuring Poverty in Peru]. *Colección Metodologías Estadísticas, 1*(2),1–8. https://www.inei.gob.pe/media/MenuRecursivo/metodologias/pobreza01.pdf

Instituto Nacional de Estadística e Informática. (2009). *Estado de la Población Peruana 2009* [Situation of the Peruvian Population 2009]. http://www.inei.gob.pe/media/MenuRecursivo/publicaciones_digitales/Est/Lib0879/libro.pdf

Instituto Nacional de Estadística e Informática. (2012). *Estado de la Población Peruana 2012* [Situation of the Peruvian Population 2012]. http://www.inei.gob.pe/media/MenuRecursivo/publicaciones_digitales/Est/Lib1032/libro.pdf

Karl, M. (1995). *Women and empowerment: Participation and decision making*. Zed Books.

Kelly, L. (2016). *The memory code: Unlocking the secrets of the lives of the ancients and the power of the mind*. Atlantic Books.

Krishnamurti, J. (1969). *Freedom from the known*. Harper One.

Krishnamurti, J. (1972). *The impossible question*. Krishnamurti Foundation.

Krishnamurti, J. (1978). The impossible question. Penguin Books.

Krishnamurti, J. (1982). *The network of thought*. Mirananda.

Krishnamurti, J. (1985). The world of peace/Welt des *F*riedens. Jadrný.

Krishnamurti, J., Bohm, D., & Shainberg, D. (1976). *The transformation of man Jiddu Krishnamurti in conversation with Dr David Bohm and Dr David Shainberg*. [Film; four video-disc DVD]. Krishnamurti Foundation.

Laplante, L. J. (2007). Women as political participants: Psychosocial postconflict recovery in Peru. *Peace and Conflict: Journal of Peace Psychology, 13*(3), 313–331.

Lindsey, C. (2001). *Women facing war. ICRC study on the impact of armed conflict on women*. International Committee of the Red Cross Women and War. http://www.icrc.org/eng/assets/files/other/icrc_002_0798_women_facing_war.pdf

Madison, S. D. (2005). *Critical ethnography: Method, ethics, and performance*. Sage Publications.

Manrique, N. (1998). The war for the central Sierra. In S. J. Stern (Ed.), *Shining and other paths: War and society in Peru, 1980–1995*. Duke University Press.

Mantilla Falcón, J. (14 May, 2002). *Informe Consultoría en Género de la Comisión de la Verdad y de la Reconciliación* [Gender Consultation Report for the Truth and Reconciliation Commission]. MEMORANDUM No 208-2002-CVR/I.

Mantilla Falcón, J. (2013). The Peruvian case: Gender and transitional justice. In L. Yarwood (Ed.), *Women and transitional justice: The experience of women as participants*. Routledge.
Nussbaum, M. C. (2000). *Women and human development: The capabilities approach*. Cambridge University Press.
Nussbaum, M. C. (2011). *Creating capabilities the human development approach*. Belknap Press of Harvard University Press.
ODE Oxford Dictionary of English. (2003). In C. Soanes & P. Hank (Eds.), *Oxford Dictionary of English*. Oxford University Press.
Pedersen, D., Kienzier, H., & Gamarra, J. (2010). Llaki and ñakary: Idioms of distress and suffering among the highland quechua in the Peruvian Andes. *Culture, Medicine and Psychiatry*, *34*, 279–300. http://www.ncbi.nlm.nih.gov/pubmed/20405314
Pomeroy, E., & Oliver, K. (2021). Action confidence as an indicator of transformative change. *Journal of Transformative Education*, *19*(1), 68–86. https://doi.org/10.1177/1541344620940815
Porter, E. J. (2007). *Peace building: Women in international perspective*. Routledge.
Rodriguez Carreon, V., & Vozniak, P. (2021). Embodied experiential learning: Cultivating inner peace in higher education. *Journal of Awareness-Based Systems Change*, *1*(2), 31–50. https://doi.org/10.47061/jabsc.v1i2.1179
Rosen, J. (2007, August 11). *Attitudes towards gender roles in the public sphere: An individual and contextual level analysis in 39 countries*. Paper presented at the annual meeting of the American Sociological Association, TBA, New York. http://citation.allacademic.com/meta/p183332_index.html
Rowlands, J. (1995). Empowerment examined. In *Development in practice* (Vol. 5, No. 2, pp. 101–107). Taylor & Francis, Ltd. http://www.jstor.org/stable/4028929?seq=2
Samman, E., & Santos, M. E. (2009) Agency and empowerment: A review of concepts, indicators and empirical evidence. In *Oxford Poverty and Human Development Initiative (OPHI)*. University of Oxford. http://www.ophi.org.uk/wp-content/uploads/OPHI-RP-10a.pdf
Sen, A. (1985, April). Well-being, agency and freedom: The Dewey lectures 1984. *The Journal of Philosophy*, *82*(4), 169–221. http://www.jstor.org/stable/pdfplus/2026184.pdf?&acceptTC=true&jpdConfirm=true and http://www.freelogy.org/w/images/d/dc/Sen85.pdf
Sen, A. (1995). *Inequality re-examined*. Oxford University Press.
Sen, A. (1999). *Development as freedom*. Oxford University Press.
Sen, A. (2000). *Why human security?* Text of presentation at the "International Symposium of Human Security". http://www.humansecurity-chs.org/activities/outreach/Sen2000.pdf
Sen, A. (2007). *Identity and violence: The illusion of destiny*. Penguin Books.
Sen, A. (2008). Violence, identity and poverty. *Journal of Peace Research*, *2008*(45), 5. Sage Publications. http://www.sagepub.com/martin3study/articles/Sen.pdf
Sen, A. (2010). *The idea of justice*. Penguin Books.
Simic, O. (2009). What remains of Srebrenica? Motherhood, transitional justice and yearning for the truth. *Journal of International Women's Studies*, *10*(4), 220–236. http://vc.bridgew.edu/jiws/vol10/iss4/1
Streeten, P. (1995). Foreword. In Haq, M. ul. (Ed.), *Reflections on human development*. Oxford University Press.
Suarez, E. B. (2011). *Surviving the "Sasachacuy Tiempu "(difficult times): The resilience of Quechua women in the aftermath of the Peruvian armed conflict*. http://idl-bnc.idrc.ca/dspace/bitstream/10625/48930/1/IDL-48930.pdf
Subramaniam, M. (2006). *The power of women's organizing: Gender, caste and class in India*. Lexington Books.
Suthanthiraraj, K., & Ayo, C. (2010). *Promoting women's participation in conflict & Post-conflict societies*. Global Action to Prevent War, NGO Working Group on Women, Peace and Security, Women's International League for Peace and Freedom.

Tadjbakhsh, S., & Chenoy, A. M. (2009). *Human security: Concepts and implications*. Routledge.

Thapa, L. (2007). *Nepal's widows*. Open Democracy: Free Thinking for the World. http://www.opendemocracy.net/blog/nepals_widows

Theidon, K. (2004). *Entre prójimos: El conflicto armado interno y la política de reconciliación en el Perú* [Between Neighbors: the Internal Armed Conflict and the Politics of Reconciliation in Peru]. Instituto de Estudios Peruanos.

Van der Kolk, B. (2014). *The body keeps the score: Brain, mind, and body in the healing of trauma*. Penguin Group.

Varela, F., Thompson, E., & Roch, E. (1993). *The embodied mind. Cognitive science and human experience*. The MIT Press.

Webb, A. K. (2011). *Nuestro propio sendero: Una comunidad andina y una economía de valores del mañana* [Our own path: An Andean community and an economy of tomorrow's values] Instituto de Estudios Peruanos.

White, S. A. (1994). The concept of participation: Transformation rhetoric to reality. In S. A. White, S. K. Nair, & A. Joseph (Eds.), *Participatory communication: Working for change and development*. Sage.

Young, I. M. (2011). *Justice and the politics of difference*. Princeton University Press.

2 *Consciousness-capabilities* under the umbrella of human development and human security

The analyses of freedom in the empowerment process of an emerging human agency need a broad and integrated perspective, a multidisciplinary study focused on the person's freedom. The two central freedoms are *freedom from want* and *freedom from fear* (UNDP United Nations Development Programme, 1994). The minimalistic definitions, respectively, are freedom from living in poverty and freedom from living in war. These two central freedoms are identified by analysing what held back the empowerment process of victims-survivors' experiences living in rural areas. The empirical data was collected to build a theory centred on the inner (what happens inwardly and at times unconsciously) and outer experiences of human agency in its empowerment process. Looking through two interdependent lenses: capabilities and consciousness. *Capabilities* means *to be* someone and being enabled *to do* things with freedom (Sen, 1999),[1] and *consciousness* is the *awareness* of our lived-experience (Varela et al., 1993, p. XVI)[2] to be free. This chapter explains the interweaving of capabilities and consciousness emerging as foundational pillars to offer new understandings of the significant challenges in human agency facing complex systems of poverty and war.

Analysis under the umbrella of human security and human development involves identifying threats to the *vital core*, which is essential for the sense of freedom in human agency. Rural people living in poverty cry out in the aftermath of war the need for development and peace with justice.[3] Capabilities as a concept based on Amartya Sen's approach serve as a "philosophical basis of human development" which aims to pursue several freedoms for development focused on the person (Alkire & Deneulin, 2009, p. 23). The capability approach and human development are known as the people-focused approach. However, human development narratives have neglected "threats" to people in conflict scenarios (Tadjbakhsh & Chenoy, 2009, p. 98). According to the UNDP Human Development Report (2005), violent conflict increases poverty, and "claims lives not just through bullets but through the erosion of human security more broadly" (UNDP United Nations Development Programme, 2005, p. 155). Therefore, an expansion of capabilities in human development emerged against the threats of wars. Human security responds to the disruption of armed conflicts.[4] Drawing on relevant philosophical key concepts of Amartya Sen on capabilities, Pakistani economist Mahbub Ul Haq introduced Human Security in the Annual UNDP Human Development Report in

DOI: 10.4324/9781003450955-3

1994 (UNDP United Nations Development Programme, 1994), "human security was not just about safety during conflict but also maintaining and safeguarding development gained" (Tadjbakhsh & Chenoy, 2009, p. 105). What is the development gained? Inner and outer development. For example, over centuries, peasants had a spiritual and scientific connection with the land, and war compromised the internal quality of its evolution.

The umbrella of human security and human development is proposed as the metaphor to cover and protect against hazards causing the bottle neck in the empowerment formation. Sabina Alkire (2003) argued that "human security takes its shape from the human being: the vital core that is to be secure", where the threat identification exercised against that vital core is necessary (p. 24). By analysing the system's human development and human security to find what are those inner threats and outer hazards to the capability to feel human and safe, we might contemplate to ask, is war preventable?

Consciousness and *capabilities* bring a base knowledge to understand the process behind statements such as, "to be lost in thoughts". A statement cried out particularly by women living in rural communities in the aftermath. Pedersen et al. (2010) noted:

> This in turn leads to the anticipation of recurrent feelings of insecurity and distress that are experienced as *pinsamentuwan* [worrying thoughts, worries], which are then attributed to consequences of *vida pobre* (impoverished life) and result in unfavorable outcomes such as *ñakary* [suffering] and *llaki* [sorrow, sadness], that is, collective and individual expressions of distress directly connected to memories of the *sassachacuy tiempu* [difficult times].

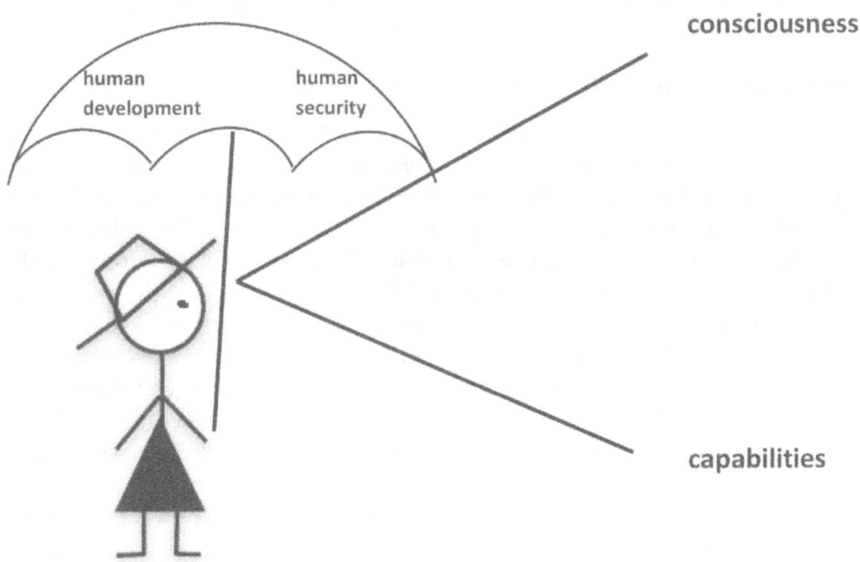

Figure 2.1 The umbrella for analysis. ©Rodriguez Carreon, V (2014).

In war, those concepts of distress are "transformed and adapted" and manifested in new experiences (Pedersen et al., 2010, p. 294). Those experiences affected the *vital core* and became an energetic obstacle for individual agency's freedom. In my fieldwork I had seen this for example in Rosario's story, when she mentioned in her testimony to the TRC, "*Sueño con mis padres, y amanezco con dolor de cabeza, en mis sueños mi madre me dice: Hija te estoy buscando*", *también escucha la voz en el lugar que mataron a su madre* ("I dream of my parents, and I wake up with headaches, in my dreams my mother said to me: daughter I am looking for you", also she heard her mother's voice in the place where she was killed).[5]

Adding the lenses of consciousness, which for now I use the minimalistic and philosophical meaning "*all* thought, feeling and action" (Krishnamurti, 1969, p. 30), enables awareness to see, feel, and perceive the blind spots. Overlooked in victims-survivors' lives during war are the freedom for awareness of their own experiences. Armed conflict threats against the freedom of having thought, feeling and action aligned with the experience of the now, by splitting the Self and its capacity to be in consciousness. Experiences like the testimony described were memories prohibited from being told in war, and they came to the surface when they were recalled in the aftermath. Much of the literature about the Peruvian conflict is loaded with stories that speak about not only what was visible, physical violence, but also what was sensed which I will call invisible violence. Invisible violence has particular inner and outer qualities manifested through marginalisation. Therefore, this study opted to consider what are all the thoughts, feelings and actions in victims-survivors that either enhance or compromise people's well-being. So, the needs of Rosario from the jungle region and the Quechua women with feelings of *ñakary* and *llaki* can be prioritised.

The capability approach: enabling the *being* and *doings* of human agency

According to Amartya Sen, pioneer of the capability approach, to enable human beings the freedom to do things or to be someone they value, and have a reason to value, is essential for capabilities (Sen, 1999). Functioning, capability, and agency are central concepts to the capability approach. "A *functioning* is being or doing what people value and have a reason to value. A *capability* is a person's freedom to enjoy various functionings - to be or do things that contribute to their well-being. *Agency* is a person's ability to pursue and realize goals she values and has a reason to value" (Alkire & Deneulin, 2009, p. 22). Understanding their interdependence is to recognise capabilities as a "reflection of freedom to achieve valuable functionings" (Sen, 1995, p. 49; Sen, 1999). Capability then is a set of functionings (Sen, 1995, p. 50), "(beings and doings) that the person can achieve" (Alkire, 2009, p. 486). As there is a set of different capabilities, for this research identifying the threat to safeguard victims-survivors' well-being aims to make visible the inner qualities that are needed for what they value to pursue. Sen's capability approach suggests a perspective based on the evaluation of the "well-being" and "freedom to pursue such well-being". Well-being is the value either for being someone or doing something. This is called the capability set, the freedom we observe to evaluate our well-being (Sen, 1995, p. 39).

Functioning is related to the achievement of well-being. For Amartya Sen (1995), functioning is the consisting of beings and doings, and the evaluation or assessment of those things is the status of our well-being. If we are doing things that we value and we have reason to value, we are achieving well-being. The achievement regardless of the freedom to achieve well-being, is called by Sen, the achievement of functioning. If we have the freedom to achieve such well-being, it means we have the capability to function. For example, Maria Guimarita (narrated in the previous chapter) had the functioning to mobilise to go to the human rights organisation. But she did not have the freedom to achieve well-being. We could have functionings without capabilities, but we cannot have capabilities without functionings; which means we can be someone and do things, and not necessarily need capabilities because we have not necessarily processed them as something we value *to do* or *to be*. "Capability is primarily a reflection of the freedom to achieve valuable functionings … In so far as functionings are constitutive of well-being, capability represents a person's freedom to achieve well-being" (Sen, 1995, p. 49).[6] In other words, the sense of value and freedom differentiates functionings and capabilities.

In involving the sense of value and freedom, the capability approach connects to the ability needed to enhance human agency. Agency is the beings and doings we value, and it has an ethical impact (Deneulin, 2006). The graphic below on well-being summarises Amartya Sen's capability approach and its analysis from Sabina Alkire and Séverine Deneulin. It shows the correlation among the concepts intrinsically reinforcing each other in relation to the need to enable people's agency.

Capabilities grounded in Aristotelian philosophy is the soil to "the roots of human development" (Nussbaum, 2010, n.d.). The United Nations Development Program analysis of the HD roots in many cultures and religions, quote Aristotle, "wealth is evidently not the good we are seeking, for it is merely useful and for the sake of something else" (UNDP United Nations Development Programme, 1994, p. 14). This suggests that wealth is perceived for the potentials and opportunities provided to the people to enhance human freedom. In this sense, human development concerns the human's opportunities as capabilities to do things or to be someone with freedom, and not merely an economic opportunity without human value. Here, I am referring to the freedom of inner capacities. Adding those in charge of social structures should desire to expand people's freedom to choose what they value and have a reason to value (Alkire & Deneulin, 2009, p. 22). Then, the status of a healthy society can be reflected when it enables opportunities for people's doings but also for people's beings to be free and to exert their potential capacities for their well-being and their community. Enabling is a critical component in the capability approach. In this process, enabling means creating the conditions to be someone and to do things grounded. Capabilities are what allows the functioning by not constraining the freedom for people to do things or to be someone they value and have reasons to value. People then can exert agency because empowerment can flow with freedom. Empowerment can be experienced if people's functionings are free to achieve the goals chosen.

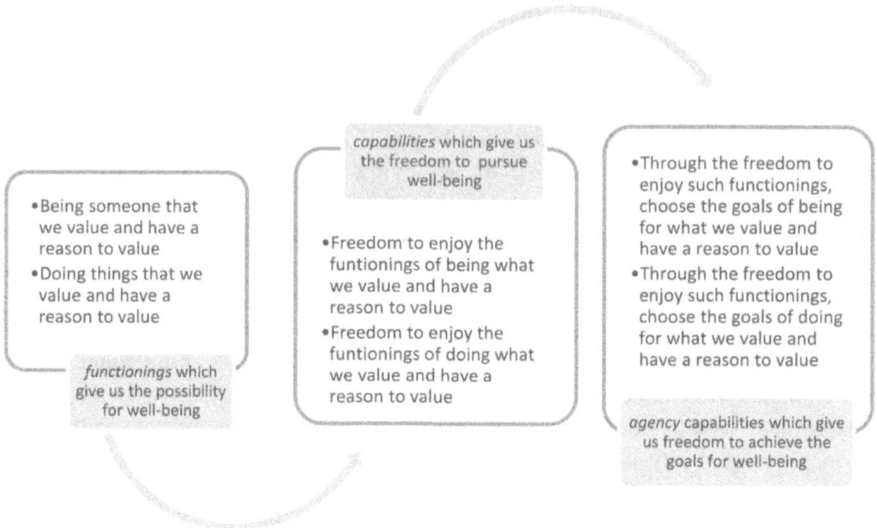

Figure 2.2 Well-being: functionings, capabilities, and agency.

Empowerment is a process where inner and outer agency evolves, and the freedom to pursue things people value flows cohesively depending on the interactive dynamic between people within their system. For example, rural women can pursue well-being based on their needs to provide food for orphans; they could act in freedom on their survival goal in the environment of Ayacucho. But urban people's inner condition of relating sympathy with rural ways of being and doings may be disconnected. Therefore, rural women agency may face different obstacles in the city, where different needs are to be met to achieve the same well-being. Séverine Deneulin analyses Amartya Sen's capability view, Sen identifies sympathy as one of the two "moral sentiments for human action to be conducive to well-being enhancement" (Deneulin, 2006, p. 9). Then, if women's agentive role is met in Ayacucho but is restricted from decisions made in Lima, their general well-being in Ayacucho is affected. Simultaneous needs interconnect and depend on each other, affecting the general well-being.

Agency, capabilities, and empowerment

Agency's inner qualities are extended when people pursue what they value, enabling the freedom for its empowerment process to evolve. Agency by itself is important as a means (Samman & Santos, 2009). Women's freedom to exercise their agency enhances women's empowerment (Ibrahim & Alkire, 2007). Therefore, the functions to be and to do are essential to well-being if there is the freedom to bear their agential power.

Agency and empowerment influence each other intrinsically and instrumentally (Samman & Santos, 2009, p. 6). Ibrahim and Alkire (2007) provided another explanation, this time through a psychological lenses, which has clear affinities with Sen's capability approach; a person is autonomous when their behaviour is experienced as willingly enacted and when he/she fully endorses the actions and/or express their values in which they are engaged. As Ryan and Deci put it, "people are therefore most autonomous when they act in accord with their authentic interest or integrated values and desires" (Ibrahim & Alkire, 2007). I argue that through sensing autonomy, being empowered (in-power), agency is embodied. It is important to note here, in Sen's approach, to be an agent with autonomy used for non-valued purposes—such as the exercise of violence—is to be a person who does barbaric things. It is not exercising agency even if he/she is challenging to be an agent. The ethical impact of the concept of capabilities is important (Deneulin, 2006). On the capability's narrative, exercising agency is an action of growth, a human growth or development on freedom from want and freedom from fear. Constructing a theory and practice that protects against the threads to the *vital core* explains why delivering barbaric actions is not a process of growth in capabilities.

Empowerment in human agency is rooted in the inner quality of being, which is different from exercising barbaric actions of collective oppression, suppression, repression, and aggression. The examples used in the case studies involved subjects such as the members of the Shining Path who massacred villages and abducted children, and the military members from special commando units who tortured and disappeared large numbers of the rural civilian population in Peru. They pursued their goals, and whether this was something "they wanted" aligned with their authentic self can be argued if we investigate the consciousness of authentic self. For the capability approach, certain concepts are associated with choices that enable well-being. Being or doings what people value and have a reason to value has a morally ethical component in capabilities. This study sees the agential characteristics to exert freedom in the empowerment process concerning well-being within poverty and armed conflict contexts.

The concept of value is applied to recognise those agents who achieve change that enhances human development. Change towards the development of the human being and differentiating from others that perpetrate structural or direct harm against the *vital core*, which happens in war. If the beings and doings threaten the connection to the sense of oneself, this attempt interferes with their freedom for empowerment to evolve. Empowerment is a process with a positive connotation, an intrinsic movement associated with freedom to be and to do. In the case of the armed conflict, committing barbaric things had reasons underneath the constrained choices that were part of functions made in an oppressed environment (for example, living in poverty) or under repressive measures used in armed conflict. For example, in Chapter 4, I write about indoctrination, and how going through it corrupts the authentic self.

In terms of capabilities, empowerment is about making choices, which may include reasonable people who make choices sometimes to be deprived in one area, and be able to be free in the area where they aim to make choices (Alkire & Deneulin, 2009, p. 36). In war, choices with freedom are conditioned systemically disowning people's ability to survive. In Peru, rural women exercised survival agency organising their private commitments, such as feeding children with other women to take turns to search for their husbands and sons by going to the police station and/or military bases. If these choices were subject to feeling coerced to give up their human dignity, in that case, women's human freedom was obstructed. In hardship environments, there are examples where people's agency is survival, and it seems passive. There are other examples where people's agency appears active but might not be conscious. They were obligated yet internalised as "choices", such as the case of a woman in the highlands who was obligated to "choose" to be raped to prevent her daughter from being raped (Theidon, 2004), and the women who was obligated to "choose" being raped to avoid their husbands from being killed. Another example occurred when the group of mothers in the Beslan school massacre near Chechnya were told to be "free" and take their nursery children and leave the older ones.[7] Such examples represent choices made by the aggressor; the "choice" was not based on freedom. It was a coercive and oppressive situation in which the connection between people's agency and its *vital core* was directly affected. The exposure to traumatic experiences forces someone to be between two "choices" against their values that lead to a torture of the psyche, body, mind and soul.

These "choices" were not based on freedom, the Self-doing split from the Self-being, because the *to be*—related to the authentic self—does not reconcile the action of doing to the being they knew. The daughter of the woman who got raped in the previous example, twenty years after, her memory of the past experience of both mother and daughter is still torturing them. All of this is blamed on the "choice that was made" twenty years ago (Theidon, 2004). The lived-experience shows these "choices" did not enable people's capabilities of freedom to pursue a meaningful life. The women in Chechnya have the crying voice of the older daughter of six years old saying, "Please, don't leave me behind" (Murphy, 2004). Her mother's experience as she said left her "dead inside" (Murphy, 2004).[8] The term "choice" implies a possibility, and in the examples given, the threat was existential. The absence of any reason consequently splits the Self and entangles with the transgressed "choice". There was no "choice" for which a new word should be elaborated when the wound is made to Self. Wounds on the inner self manifested through a torture to the thinking, and to the sense of being in this world are hazards from war. The threats from war are not only the death of physical bodies but the death of the spiritual agent that lives within the human. It threatens a collective disconnection with the *vital core* of human security. The latest Human Development Report (2021/2022) shows what happens when there is no psychological resilience and human development in capabilities. How mental well-being can influence choices and behaviour at multiple stages of functioning, affecting the ability to transform what is valuable into capabilities (UNDP United Nations Development Programme, 2021/2022, p. 75). Therefore, to exert agency, the freedom capabilities in those cases were perceived as non-existent because functionings were repressed.

Consciousness analysis of the inner experience is necessary to understand the capabilities of freedom in relation to how people's agency is processed. People in those examples did not have choice and the ability to exert agency, because "the opposite to a person with agency is someone who is forced, oppressed, or passive" (Alkire & Deneulin, 2009, p. 37). My focus is on those conflicted experiences embodied by victims-survivors' agency. The wounds from the war threatened their communities with a collective disconnection from their *vital core*. Their karmic condition is in the collective field; for neuroscientist Francisco Varela et al. (1993), karma as a process is "the main condition in the accumulation of conditioned human experience" (p. 116). Depending on how the vital core has been conditioned intrinsically, the agency's enabler is, in either case, numb (inaccessible) or moving (evolving), though people cannot see it cognitively. The language to understand the inner capabilities of agency requires us to tap into different levels of embodiment knowledge.

Core value: the *vital core* and motivations

Empowerment transforms intrinsically as an extension of agency. The dynamic of agency within the empowerment process of formation and/or transformation is conditioned by motivation for participation and decision-making. If in times of conflict, the freedom for functioning is constrained, it does not necessarily mean that people stop doing things or being someone. As long as the *vital core*—the sense of oneself—is there, then agency can still evolve in its empowerment process with motivation. Motivation enhances well-being within the *vital core* connection of human agency. Motivations alone, according to Herbert Petri (1986), are forces acting to "initiate" and "direct" something, and describe the intensity and direction of the behaviour (p. 3). Well-being permits us to evaluate a stage of agency when empowerment is forming and/or transforming. Following Sen's argument that a person cannot be an agent if what he/she pursues is not based on motivation for human development's well-being yet to perpetrate harm, this intersectionality leads us to raise awareness of the threat to people's *vital core*. It is necessary to consider that people's vulnerabilities are higher in armed conflict, and the pursuit of well-being is driven by different motivations. Moreover, motivations are intensified to satisfy the need to escape that existential vulnerability. Without paying attention to consciousness, the safeguard to the sense of Self risks the owning of one's vulnerability.

The *vital core* and the motivation to protect it become the core value for agency to be pursued. It is in this action (of pursuing) that the agent is defined. For example, the core value for the late Mendoza de Ascarza might be the love for her son, and that is her primary agency to pursue justice (doing) that became a need after her son's disappearance. The injustice she had experienced became the motivational seed for her empowerment process to evolve. As long as the motivation—the seed of justice—was there connected, she continued searching her son. The core value that defines her as agent, in this case love that comes from her motherhood capability, is what motivates her empowerment transformation. The core value is an internal force that has an internal goal, which is in her case, to feel inner peace by finding the whereabouts of her son.

In this case study, justice for many Quechua women in Ayacucho became a "thing to do and pursue" and had different motivations among the different rural women. Sen (1995) argues that freedom for beings and doings what we value means that we are not constrained in our decisions or choices. However, in the case of rural women, the "choice" was made within a controlled and dangerous environment, as poverty and armed conflict threatened rural women's capacity to seek justice. Therefore, the motivation emerged from the problem of injustice. The core reason for that motivation did not emerge from free choice; it was the sense of injustice that became the reason for the motivation.[9]

I here underpin the analysis of motivations and the choices—which can be made with constraints or with freedom—to achieve needs. For rural women in an underdeveloped context the need is to get out of poverty. According to Johan Galtung (see more, Galtung, 1990) poverty is a structural violence, "occurs when social structures undermine individual wellbeing" (Navas et al., 2018, p. 651). It is a context where choices are not made in complete freedom. The pursuit is the motivation "to not give up"[10] towards the path of development. In armed conflict, choices were motivated by the lack of peace, for example, the need to escape direct violence. Within an environment of poverty, the appearance of armed conflict intensified vulnerabilities and deepened survival agency. As Stewart argues, most conflict occurs where poverty is already a setting (Stewart, 2002, p. 105) and increases in both the number of violent acts, and intensity of violent cruelty. Particularly in towns isolated from where it can be seen and accounted for. Conflict exacerbates poverty and adds complexity to the meaning of peace, with various dynamics involving different needs that need to be met. All of them are interdependent and affect each other's motivations to achieve needs. Therefore, when there is more than one "thing" to pursue that is perceived as equally valuable, motivations to make choices overlap between free choices and constrained choices. In this dynamic of interdependent choices, well-being and, consequently, the *vital core* can be affected.

The *vital core*: the compromised functioning

Here, I focused more intensely on identifying the threat to the *vital core* of human experiences. Much of the literature that emerged from the 20-year armed conflict in Peru, like other wars, is loaded with survivors' dehumanising experiences (Comisión de Entrega de la Comisión de la Verdad y Reconciliación [CVR], 2004). For example, the story collected by Kimberly Theidon (2004) about Felicitas and her narrative of the *llakis*.[11] *Llaki* is a Quechua word to express the distress of sensing sorrow and sadness (Pedersen et al., 2010). Felicitas describes having seen her husband being killed, he was tied to a mule and dragged along the ground before he was knifed. She and her sons saw everything. After that, they felt persecuted by both terrorists and military members. She remembered that they couldn't stop thinking. They were trapped in their thinking, as they said, they were *crazy*, they were with *llakis*, and they died with *llakis*. The *llakis* is defined as one of the principal ills (Theidon, 2004, pp. 64–66).[12] *Llaki* is also a particular name for psychological trauma among the Quechua people (Watters, 2010, p. 103; Pedersen et al., 2010).

In the case of Felicitas, her mental well-being was affected, and her *vital core* compromised, obstructing her capability *to be*. Her sense of agency was disconnected from her *vital core* because the *llaki* "took over" her reason to value her sense of living. It is important to understand the *vital core* as an analysis of subjectivity. Jhon M. Alexander (2008) does a comparison of the capabilities' definition between Sen A., Nussbaum and Anderson. His emphasis is that Nussbaum claims some capabilities in her list that are abstract and, therefore, hard to implement (2008, p. 66). However, analysis of choices for pursuing beings and doings based on abstract motivations should not be ignored. In particular, because abstract motivations are transforming agency inwardly. The phenomenon between the *vital core* and agency is "visible" when observing the quality of the connection between the inner and external Self. Such qualities enable to be in a state of being or manifest the to do in a particular way. What is "invisible" to cognitive reasoning is the inward phenomenon happening and which is capable of being perceived by sensing.

When applying the capability approach, the analysis of freedom in armed conflict alone cannot be seen without the constraints emerging from living in poverty. They are interdependent as a context. In a rural village where people did not have access to schools, after the war, electricity and roads were presented as things that the State should have provided. When it came to reparations, victims-survivors prioritised them as development. For example, schools built as part of the "repair" of the unjust experiences that were the consequences of armed conflict. Repairing the visible (items, infrastructure, or spaces for services) for unmet needs before the conflict would have been seen as a sign of development. To repair with the "visible" to violations against the inner self (massacres, disappearing, or torture) experienced during conflict is problematic. I argue the need in the village to attend the *llaki*, and consequent support for their mental well-being could have provided access to human development policies for people living in poverty. To focus on the *vital core* means to work towards integration and healing the traumatic experiences that they had survived in war. A domino effect can be seen from the effects of violence. For example, school attendance might be threatened when *llaki* needs are not attended to because the children's concentration was affected. In other words, it is like sending children to school without food or with a sleep deprivation. The children may want to study (doing) or be good students (being); however, these obstacles carried in their unconscious will result in constraining them from exercising their capabilities with freedom. Being in freedom is to be connected to the authentic being.

The emergence of human development and human security

Traditional development and security focus on poverty and conflict, respectively. They are not the focus of this study in identifying threats against the empowerment process because human agency and well-being are not at the core of their approaches. According to Martha Nussbaum (2011, p. 1), traditional development as an economic measure of gross domestic product (GDP) did not recognise or constitute the human focus of development programs, like human development policies. In other words, GDP does not necessarily make a difference in the quality of people's lives. Mahbub ul Haq (1995, p. 27) adds that "the mainstream of development ... regarded as a residual of economic growth, ... is inconsistent with human development strategies—which are focused on investment in all the people on their full participation in human well-being". Meanwhile, traditional security focuses on a military source of solutions to conflict and state affairs (rather than on the people), a type of security that holds back people's empowerment in war. Deneulin (2009) argues that the military body has long relied on security as its key concept, based on the increase in conflicts in many developing countries. Poverty as an injustice concept has been seen as a possible cause of conflicts in which minimising inequalities and promoting development have been seen as a solution for security and peace, and therefore justice. However, military intervention is not justified for addressing insecurities within the human security approach. On the contrary, "the insecurities that threaten human survival or the safety of daily life, or imperil the natural dignity of men and women, or expose human beings to the uncertainty of disease and pestilence, or subject vulnerable people to abrupt penury related to economic downturns, demand that special attention be paid to the dangers of sudden deprivation" (Deneulin, 2009, pp. 62–64). Traditional security focusing on military intervention does not enable the freedom of human agency and is inconsistent with the strategies that work to abolish poverty and armed conflict.

The freedom from the consequences of living in poverty and the freedom from living in armed conflict is the centre of human security. Through protecting and empowering the people to endure hardship, human security attempts to overcome uncertainty caused by deprivations (Deneulin, 2009, pp. 62–64). A system achieves focus on human freedom when "under human security, the state shifts from militarist concerns towards empowerment of its people" (Tadjbakhsh & Chenoy, 2009, p. 132). The concept of freedom in human security is fundamental for empowerment processes. The Commission on Human Security states that, "Human security naturally connects several kinds of freedom—such as freedom from want and freedom from fear, as well as freedom to take action on one's own behalf" (2003, p. 10). Those freedoms are necessary to exert the agentive role by closing the decision-making power gap; the "who" that makes decisions needs to be closer to the "receiver" affected by the decision's well-being.

Freedom from want and freedom from fear

Under the human security umbrella, freedom from want is the need for people living in poverty, and freedom from fear is the need for people living in war. These two freedoms are recognised by the United Nations (UNDP United Nations Development Programme, 1994, p. 3; Commission on Human Security, 2003). Freedom from want means "gainful employment, food and health", and freedom from fear is the absence of "physical, sexual or psychological abuse, violence persecution or death" (Tadjbakhsh & Chenoy, 2009, p. 24, 39). Freedom from fear also responds to security between and within nations. The UN compromise as a "socio-economic development, to peace within nations" is a challenge that Mahbub Ul Haq called a "supreme challenge" for our world (1995, p. 43) to address the two freedoms. Freedom from want represents no longer living in poverty or lack of human development, and freedom from fear represents both, but means no longer living in armed conflict or lacking human security. Human security connects several freedoms (Commission on Human Security, 2003). An example with a graphic illustration of how these freedoms interrelate is below.

The human development approach suggests different dynamics for development policies ensuring the development of people. I mentioned that the focus on security aspects in the case of conflicts with abrupt threats is vague and is insufficient to explain sudden threats, such as those that happen in war. Human security emerged from the need to address the threats from conflicts and others—such as economic crisis, "sudden destitution", as Sen explains, and natural disasters (Tadjbakhsh & Chenoy, 2009, pp. 98–99). Haq suggests that human security is a concept emerging from the people and their needs in response to these sudden threats (1995). The overlooked sudden threat is the need for awareness of the inner development rupture. Being conscious complements to the understanding of how the lived-experience is free.

The people-focused approach: human development

The human development approach is focused on the human first and provides a framework for analysing vulnerabilities. This is an approach that "puts people first", and the aim is to extend what people can do and be (Alkire & Deneulin, 2009, p. 23). Mahbub Ul Haq (1995) pioneer on human development theory, said,

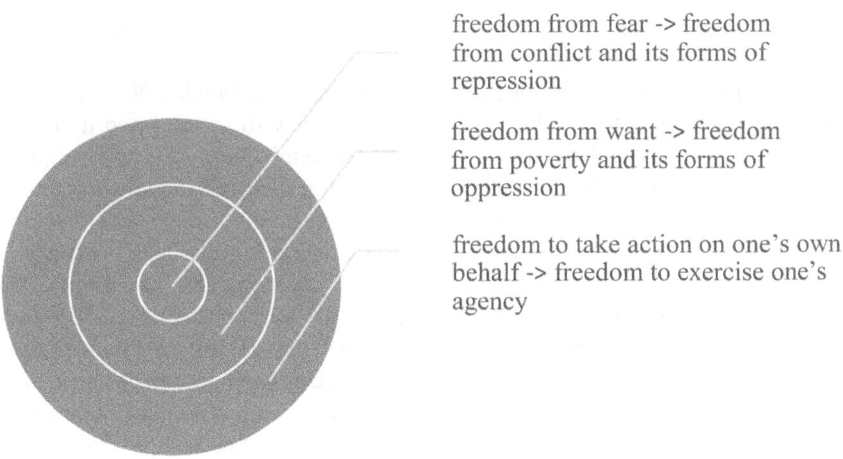

Figure 2.3 Freedom for human agency in war and poverty. ©Rodriguez Carreon, V. (2014).

"the heart of this concept is equity" of opportunities, not necessarily results, but the process whose end is human welfare (p. 117). The human development approach focuses on people as the centre of development by enlarging human choices. Poverty in human development means that opportunities for human beings are constrained (Haq, 1995).

In Peru, according to the TRC, most of the people living in poverty and armed conflict had particular identity features. They were rural and peasant (CVR, 2004, p. 22). Rural people's experiences in poverty equal limited opportunities, and when adding the phenomenon of repression, they were also characterised by being powerless to exert agency. Repression is contrary to the concept of empowerment, where "people are in a position to exercise choices of their own free will" (Haq, 1995, p. 20). In poverty, there are barriers to choices and opportunities for rural Quechua people. In war, the same system with the same consciousness controls the lack of choices and opportunities. With consciousness, human development offers hope to enable opportunities for people living in vulnerable contexts.

People living in poverty had not experienced State development, and in war, their vulnerability increased, *survival agency* became more visible, and choices and opportunities depleted. Pilar Coll (personal communication, 2009) said that development (referring to State development) is unknown or an illusion for people who have never experienced it, like the rural Quechua women in the isolated communities of the highlands of Peru. Then, after conflict, there is an increase in motivation to access development that enhances the empowerment process (Kieffer, 1984). Or, as John Paul Lederach argues, the need is for "moral imagination"—for this study in development, to overcome difficulties in "human affairs". He presents how a few women in Wajir (Kenya) got together with the determination to stop war because they wanted to be able to go to the markets without the inabilities that fearing violence brings. These Kenyan mothers co-created a safer market and constructively found new ways to engage people in peace negotiations through critical dialogues to overcome their obstacles. Poverty was still a significant challenge, but the women found a safer way to go to the market and, consequently, to continue their development (Lederach, 2005, pp. 10–19). Defeating challenges by raising awareness of agency to create opportunities for new outcomes is what Freire (1973) called conscientization. The opportunity for people's development arises when there is a shift in their motivations to construct an environment that enables them to free themselves.

Living in poverty and armed conflict are the vulnerable grounds of deprivation in accessing decision-making. As mentioned before, the motivation to get out of poverty is development, and to get out of war is peace. However, the reasons for endurance and survival overlap between peace, development, and justice. Frances Stewart (n.d.) stated that "lack of [Human Development] HD is an important cause of war … [and] anyone concerned with HD MUST, [do two things], consider policies DURING conflict [and] consider policies towards prevention of conflict". If the aim is to prevent war, attention is necessary towards those under constant hazards of vulnerable contexts such as oppression, and repression.

Vital core: the vital need for human security

The importance of analysing an armed conflict under the umbrella of human security arises from noticing that victims-survivors felt fearful and insecure, and these feelings were manifested in idioms of distress such as *llaki*. Examples of Rosario or Felicitas' (Theidon, 2004) experiences are manifested in different languages, not only Quechua, Aymara, or Spanish, but also in their own expressiveness and uniqueness of consciousness. The UNDP United Nations Development Programme (1994) argued that people's anguish comes from feeling insecure about everyday problems. The women raped, the crime, the threat of pollution, the despair from terrorism, and circumstances without sense make people embodied to feel insecure. "A people's concept and a people's concern, human security is reflected in the shrivelled faces of innocent children, in the anguish of the homeless, in the constant fear of the jobless, in the silent despair of those without hope" (Haq, 1995, pp. 116–117). In Haq's statement, the sense of despair of those people not sufficiently protected or able to change their situation, raises the need for a new human-centred approach focused on safety.

Human security protects people's vital freedoms, ensures they are protected from "critical" and "pervasive" threats and at the same time constructs an environment for people's "strengths" and "aspirations" to be fulfilled (Commission on Human Security, 2003, p. 1). When applying the human security approach, it is necessary to support the empowerment of individuals and their communities (Tadjbakhsh & Chenoy, 2009, p. 134). Collaborating with women's empowerment for engendering policies in governance, allows the governments to see women's perspective for peace and promote their essential participation to human security (Porter, 2007, p. 12, 35). "Protection" and "empowerment" are necessary for an environment with human security. The first safeguard protects people from "danger", and the second "enables people to develop their full potential and become full participants in decision-making".[13] According to the Commission on Human Security:

> [Human security is] to protect the vital core of all human lives in ways that enhance human freedoms and human fulfilment. Human security means protecting fundamental freedoms—freedoms that are the essence of life ... The vital core of life is a set of elementary rights and freedoms people enjoy. What people consider to be "vital"—what they consider to be "of the essence of life" and "crucially important"—varies across individuals and societies.
>
> (Commission on Human Security, 2003, p. 4)

I am arguing that the *vital core* is the sense of Self. The quality of being able to sense oneself enables human agency to evolve, and human security should guarantee its freedom to develop. According to Haq, an environment or system where the people are acknowledged is contrary to a society that overlooks its citizens' human security (Haq, 1995). The importance of human security lies in looking more closely at the concept of motivations as the main needs of people. To incorporate the concept of motivation and need within the human security discourse, I have applied Alkire's "threat-identification exercise". The satisfaction of that core need is intimately related to the central concept of human security (Alkire, 2003, p. 24). I have called it the *vital need.* To obtain human freedom, the *vital need* as an extension of the *vital core* identifies threats related to human safety. Alkire's identification means "the vital core that is to be secure" to have "a minimal or basic or fundamental set of functions related to survival, livelihood and dignity". Furthermore, the environment or system in place to protect human security may not provide every part of the well-being, but at the least, the core to ensure survival is met (Alkire, 2003, p. 24). *Vital core*'s protection is beyond physical, it is to protect the dignity of the human condition, and the possibility of threats is greater for those living in poverty as it increases their experience of living and sensing a non-human condition. The threat identification to human dignity requires consciousness to have an awareness of what is happening to the Self inwardly. It has sensations with inner qualities invisible to the cognitive reason and yet noticed with the heart's intelligence.

"The human security threat analysis" (see more, Tadjbakhsh & Chenoy, 2009, p. 17) mainly identifies direct threats against *vital needs*. For victims-survivors in the aftermath, threats against their empowerment process in their access to justice were critical. Alkire (2010) writes that human security focuses on the minimum set of capabilities, "protecting vital capabilities from critical pervasive threats" (Alkire, p. 60). *Vital needs* protect from deprivations when living in poverty, which comprises, as stated by Haq and Alkire, "the worst aspects of human deprivation (Haq, 1995, p. 185; Alkire, 2010, p. 30). Therefore, threat-identification is necessary for the protection and empowerment of people. As stated by the UN Office of the Special Adviser on Gender Issues and Advancement of Women Department of Economic and Social Affairs (1999, n.d.):

> We empower women in all the dimensions of the human security agenda. In other words, we are faced with two tasks. First we must identify those threats to women's security that affect them because of gender factors and second, we must seek to outline ways in which we can work so that women can be empowered and assist themselves.

To build "sustainable peace", during conflict and in the aftermath, policies and international programs are implemented (Norwegian Ministry of Foreign Affairs 2004, p. 5; UNICEF United Nations Children's Fund, 2011, p. 14).[14] The call is for sustainability and for consideration of those under the disasters of poverty and armed conflict risking being deprived of their capacity to sense themselves. The strategy of human security in peacebuilding settings is to be used as a sustainable and preventive mechanism against hazards that enhance people's empowerment and threaten their vital freedoms. If a country's system is based on human security, then it can identify threats (Tadjbakhsh & Chenoy, 2009, p. 39). In conflictual times, people's needs require a strategy based on human security concerns that attends to *capabilities* and *consciousness*.

Agency's capabilities are needed to overcome the feeling of insecurity in times of hardship. "The sense of agency is the internal aspect of feeling capable of acting upon one's goals" (Strandberg, 2001, p. 5). The sense of people's agency is the exercise of capacities in what they value and believe. In systems, Haq argues, "the obvious is the most difficult to see: The true wealth of a country is its people" (Tadjbakhsh & Chenoy, 2009, p. 104; see also, Alkire, 2010). What are the threats to institutions that blind them to see their true resources? To have a panoramic view, approaches proposing subjectivity and objectivity are required to "get an accurate picture of empowerment processes" (Strandberg, 2001, p. 7). A holistic human security approach, including the inner and outer qualities of people and institutions guarantees the exercise of people's agencies by locating the threats to prevent, and enabling people's opportunity for freedom to be empowered.

Awareness of the experience: the concept of consciousness

For people living under oppression or repression, hardship can mean a life-or-death situation. "[The people of] Ayacucho construct and experience expressions of distress and suffering such as *pinsamientuwan* (worrying thoughts, worries), *ñakary* (suffering) and *llaki* (sorrow, sadness), in a context of persistent social inequalities, social exclusion and a recent history of political violence" (Pedersen et al., 2010, p. 279). *Llakis,* as an illness from the state of thoughts (Theidon, 2004), expresses the sickness taking form from exposure to armed conflict, witnessing the deaths of family members, and chronic living under extreme deprivation of different needs. To enable functionings and connect with the *vital core*, healing capabilities are needed, different to Western psychology. Rural people had a different relation to engage with their idioms of distress.

In hardship, opportunities for people to become conscious are heightened. Paulo Freire writes that in times of transition, opportunities are raised for the oppressed to become *conscient*. They can be responsible for their decisions without being conditioned and become conscientised through their own experience. This transition of "consciousness makes the man 'permeable'" (Freire, 1973, p. 14), which means being ready to understand the self and his/her surroundings without judgement and resistance to engagement of comprehension. For example, rural people became more aware of their deprivations. For communities, being conscient is the experience of engaging with the collective existence, their reality. "Existence is a dynamic concept, implying eternal dialogue between man and man, between man and the world, between man and his Creator. It is this dialogue which makes man a historical being" (Freire, 1973, p. 14). The deep understanding of judgement comes from this internal dialogue about the relation with the Self, the Other, and the Cosmos/Environment. Freire calls this the process of "*conscientização* [which] represents the *development* of the awakening of critical awareness" (1973, p. 19). Conscientisation, or raising consciousness, can help people under circumstances of hardship to become aware of their lived experience, to be free.

The concept of consciousness assists in understanding what happens with someone's belief system when experiencing sudden traumatic events during conflict. As Friedemann Schaub (2012) writes, this belief holds back empowerment the moment it becomes part of people's identity, yet by identifying the real essence in consciousness, healing can happen. What Schaub (2012) refers to as essence is referred here as *vital core*. Freedom from incoherence is the awareness of the experience. The concept of consciousness is to evaluate the capabilities needed for the empowerment process to evolve from conditioned autonomy. Autonomy is what Freire called the process of "naïve consciousness". This means someone characterised by "over-simplification of problems", always thinking about the past without motivation to know, or underestimating "the common man". The common person, for Freire, may not attain complete consciousness. He gives examples of peasants in times of transitions (Freire, 1973). Freire emphasises raising consciousness rather than being conditioned to act by circumstances. Work on consciousness supports individual and collective alchemical agency processes to transform the system/environment and safeguard the *vital core* for human development.

The vital core: *one* affects the *wholeness*

Attachment has a key meaning in understanding the freedom of the *vital core*. Krishnamurti (1972) said that attachment is holding on to something, when "the mind is holding on to memory" (p. 115). Attachment in conflict fragments the relationship with our *vital core* affecting the wholeness of our perception, our sense of what the world means. Because attachments produce what Krishnamurti calls "fragmentation of the thought", when we are attached to one thought we exclude other thoughts and that provokes conflict. In fragmentation, we cannot see the wholeness. To analyse the relationship among the concepts I quote Krishnamurti on how attachments are related to fragmentation:

> When we face our innumerable problems we are inclined to try to solve each problem by itself ... with the problem of violence or starvation, which we try to solve politically, economically or socially. I wonder why we try to solve each problem by itself. The world is ridden with violence; the various powers that be try to solve each problem as though it were apart from the rest of life. We do not consider these problems as a whole, seeing each problem related to other problems and not in isolation... I think this must be understood ... there are thousands of problems which appear to be separate, which we never seem to see as interrelated, but no problem can be solved in isolation by itself ... do understand how all problems are interrelated ... If we want to understand the problem of violence completely, totally, so that the mind is altogether free from violence, I can only understand it by understanding the structure of thought ... 'my' house, 'my' wife, 'my' country, 'my' belief, which is utter nonsense ... The 'me' is brought about through thought; *it has no reality by itself*. The 'me' ..., 'my' desire to become somebody – that 'me' is the product of thought which divides.

> When the mind works without 'knowledge', experience then the whole of life can be observed without the 'me'.
>
> (Krishnamurti, 1972, pp. 41–49)

The analysis of fragmentation by Krishnamurti calls for awareness to see the parts as part of the wholeness—to see each experience as part of the total reality. He said that people attached to one event fragment the whole experience, and then they attempt to resolve the event without its wholeness. Applying to the war context, Krishnamurti's lenses of fragmentation and Freire's analysis of beliefs, we find that fragmentation heightens our attachment to one belief, which, as a result, becomes the attachment to one narrative that represents the individual or the collective fragmented perception, all that can be seen. In contrast, Capra brings the concept "conscious experience", which is an eminently integrated experience from different events. In his scientific view of research, conscious experience occurs "when the activities of different brain areas are integrated during a brief moment … and at other times not be part of it [dynamic] and thus be involved in unconscious processes" (Capra, 2002, p. 51). The problem then lies in looking, for instance, at the aftermath of conflict, at women's experience of oppression from poverty as one dimension and women's experience of abuse in rape during war as another issue. The experience of rural women should not be seen as isolated. Consciousness does not compartmentalise experience. Examining the aftermath to analyse oppression and aggression as interdependent issues is a step forward to understanding the complexity of what is happening inwardly. The inner self also is space for a multidimensional architecture of experiences. Friedemann said, the changes in the inner conflict belong only to the person who had the experience (Schaub, 2012).[15] For rural people to own their internal experience and what happens with their emotions is fundamental. However, through policies, rural women's experiences were categorised and fragmented into concepts, with an outer rational label for their inner embodied experiences. Experience can be a brief moment, but what happened in that moment is always represented in memory; it is "embodied", storing specific emotions, and an "embedded" perception that will shade the complete experience (Capra, 2002, p. 50). If attachment to a particular experience is reproduced through memory—a traumatic memory during the war, for example—then the freedom of the *vital core* is compromised. Adding to this, when people are unaware of what is happening inwardly, they see memory as something separate; the person experiences symptoms, as Quechua women did with idioms of distress expressing *llakis.* Judith Herman brings another way of defining idioms of distress in Western literature, the concept of trauma, which is when the person "loses her basic sense of self" (Herman, 1992, p. 52). A contrary effect of a "panoramic perspective … [is] called awareness" (Varela et al., 1993, p. 26). The person becomes attached to the memory (Krishnamurti, 1972, p. 115). In the armed conflict literature, trauma is a narrative not seen as part of the victims-survivors multidimensional stories. However, the *llaki* word has an extended meaning of trauma as Andean cosmology, the individual is seen in relation to others and their System/Environment. Because *llaki* challenged Andean life, trauma for this research includes the lenses of *llaki* and its relationality.

Victims-survivors living during conflict experienced several traumatic experiences, which affected their thoughts. For Herman (1992), the traumatic experience forces the "survivor" to keep in mind early "struggles" with the function of control over their own "autonomy, initiative, competence, identity and intimacy" (pp. 52–53). I mentioned that *llaki*, or sickness of thoughts, occurs when the brain cannot stop thinking about traumatic experiences. In Quechua culture, this means the person cannot function in communal life. It affects their *vital need* to be part of the community. A craziness status where thoughts just hold into at any time, and you cannot get rid of them (Theidon, 2004). It is a stage that troubles the mind. Felicitas and most victims-survivors who experienced a *llaki* stage want peace in their minds, which means the healing of their wounds that blinded their capability to see what they valued and have a reason to valued.

The trauma fragmented from the wholeness becomes attached to a particular time, in this case, the war. An individual and/or collective experience during the confusion of war affects the *vital core* and unconsciously distorts the consciousness of the empowerment process. Herman argues that traumatic experiences create "disempowerment and disconnection", where the survivor must become the author of his or her own change for recovery (Herman, 1992, p. 134). Thich Nhat Hanh said this recovery to peace of mind would not come just from eliminating violence, but nurturing understanding of the thoughts by being aware of them and their consequences (Thich Nhat Hanh, 2003, p. 5). "An examination of experience with mindfulness/awareness reveals that one's experience is discontinuous—a moment of consciousness arises, appears to dwell for an instant, then vanishes, to be replaced by the next moment" (Varela et al., 1993, p. 73). When being aware of the fragmentation when the split in the experience is generated, awareness alone opens the possibility of becoming conscious of the effects that trauma caused, opening space to enable new inner experiences to create new feelings. As previously mentioned, during transitions, such momentum for awareness is more likely to appear (see more, Paulo Freire, 1973), including also when mechanisms of transitional justice are established. However, it is not just opportunities that arise in transition; it is also the symptoms of the wound of what happened. For example, the late Angelica described that she became perdida en (mi) pensamientos—"lost in (my) thoughts" (personal communication, 2009), when speaking about the scar that she carried.

The same consciousness that creates institutionalism organises transitions and their processes, causing the fragmentation of such an otherwise interconnected systems. Institutionalism divides human experience to understand how to support first, and at times, how to control them to get the support. Then, understanding becomes the central need for the institutions to be in the known, and the need that created the support becomes a transaction (beyond the action) to control. In that way, the proceeding is known. Understanding, however, requires a different inner quality, a dynamic to be comfortable with the unknown. For example, fragmentation within institutionalism in Peru occurred when the TRC launched enquiries about women's traumatic memories to know what happened to their loved ones. Through navigating the institutions' language, those who became part of the system wanted to find the truth and created a mechanism for their understanding of that truth.[16] Understanding, however, requires trusting the witnessing of truth vs projecting an understanding of the truth. Rural victims-survivors experiences were mechanically contextualised by compartmentalising them into known concepts. The structuring of these layers through conceptualisation is intertwined with the process of executing the label's expectations and distancing the victims-survivors' truth-seeking from their inner experience, which at the same time, is the source of their original truth. When human experiences are separated into concepts, the person no longer owns the experience. To heal, it needs to re-own the agency of narrative "given" to, for example, the TRC inquiry. This dynamic, once the lived-experience is stirred and then fragmented into several small processes to access the truth, affects the consciousness, which is what Krishnamurti (1972) called the sense of totality. The mechanism of institutionalism to exercise justice in transition happens in public hearings, registration of victims, testimonies, reparations, and other transitional processes.

Victims-survivors are told to articulate the story when the raw, incoherent, or what Herman called unspeakable truth (1992) is more truthful. Experiences like this cause what Campbell calls, in her "Post It Notes Analogy for Effects Of Trauma On Memory", fragments scattered all over the place; the experience is not altogether in order (SASC, 2014). In public hearings, for example, whenever the victims-survivors were asked to divide their experience into separate matters to recall the experience, several connections within the experiences were blocked and disconnected due to the *vital core* being compromised. Fragmenting the experience to fit a structure disconnects from the story. Truth is the witnessed incoherence needed to become conscious to be wholeness.

The interdependence of the key concepts in consciousness

According to Krishnamurti, consciousness and awareness are the same (1972). At the same time, consciousness demands the conjunction of both "body" and "mind" (Varela et al., 1993, p. 113). In other languages, such as Tibetan, awareness is translated as the "inner experience", as the "mind" which includes emotions and feelings (Dalai Lama, 2013, p. 24). Freire's (1973) known raising consciousness approach is a call for awareness. For Krishnamurti (1982), we are not individuals in consciousness because we are our consciousness, and its content is "the common ground of all humanity". When we become aware of that, the experience becomes the person and not a matter that is separate from the person. The person can become aware of what happened in the experience without judgement, and it becomes a conscious experience. Capra (2002) adds that a conscious experience of our existence *is* consciousness (p. 38). In that sense, I, for example, become aware that people's experiences are not a separate matter, but also, they are part of me.

Another key concept to understand consciousness is the meaning of thoughts. What is the relationship between consciousness and thoughts? Consciousness is an interaction of doings and beings through the process of thinking. For Bohm, "thought is doing it. But doesn't even know it is happening" (Bohm, 1992; Bohm, 1994, p. 116). Krishnamurti explains how thoughts connect with memory:

> Thought is the response of the memory of things past; it also projects itself as hope into the future. Memory is knowledge; knowledge is memory of experience. That is, there is experience; from experience there is knowledge as memory; and from memory you act. From that action you learn, which is further knowledge. So we live in this cycle – experience, memory, knowledge, thought and thence action – always living within the field of knowledge.
>
> (Krishnamurti, 1982, p. 88)

When applying to the analysis of war, consciousness is the noticing of the happening, a process which is extremely difficult for people living in the confusion of conflict. David Bohm emphasised the implications of conditioning. Written after his death by Sarah Bohm, he said, "thought works by conditioning. It has to get conditioned. You need conditioning to learn a language, to learn how to write, or to do all sorts of things". Everything can be conditioned, but if it is not rigid, we can reflect upon those conditions. But if the conditioning is too rigid, "it won't change" (1994, p. 120). In the case of the people living in poverty and its forms of oppression during armed conflict, thoughts become conditioned by their environment living in violence, in that context, they are embodying their beings and doings.

For example, Freire noticed that "the thought-habits of peasants" are so close to the "natural world" that they feel more part of this world than transformers of this world. This interaction between the peasants and their environment made it a difficult transition to "enter into" the world as subjects of change. Because to enter the world, it is necessary to "becom[e] aware of reality", which means for the peasants to have a feeling of belonging to an experience of reality (Freire, 1973, pp. 105–106). Since colonialism, in 1526, Quechua people in Peru experienced oppression. It could be argued that they feel more part of what it means to live under oppression rather than becoming an observer of themselves during such oppression. For Freire, for the peasants to observe and become aware, they must capture the relationships between "distinct perception" and other "dimensions of reality". This transformation is possible with *conscientização* (conscientisation, or raising consciousness) (1973, p. 110). For this to be possible, awareness of the attachments, to the meanings of oppression, for example, is necessary as an embodied experience. Freire mentioned that "detaching themselves from their surroundings, they transform their environment" (Freire, 1973, p. 111). This does not mean they detach from nature; it means that peasants must become observers of their experience and their environment because they are both, the experience and the environment. Once they have the freedom to embody, they can transform their experience and their environment. When I refer to the concept of environment, it is not nature alone but the systems and cultures and include the relationship as how people become part of, different from Environment, which I relate to the Earth.

The process of becoming integrated with experience is to be aware of the thinking process. In the process of thinking described in Krishnamurti's narrative, experience is the beginning, and it reproduces a new experience that is simultaneously conditioned from the past memory, the knowledge, the thoughts, the responses, and the reactions that emerge from the new knowledge and action.

Awareness of the process is consciousness, in which the *vital core* can only be free if there is no attachment to one of the fragmented parts of this process. This, however, will depend on how much the person is conditioned to the dynamics of violence. If the person's being and doings were conditioned by experiencing oppression, the process of detachment will be more difficult if, as a consequence, their *vital core* was compromised. In other words, the person is conditioned by what they know to do and to be. Nevertheless, for Freire and Krishnamurti, an awareness that arises in transition may enable the freedom of the person's agency as part of their own empowerment process. As a result of this transformation, the person's agency can be exercised, and empowerment enhanced.

Conclusion: merging the theories of the people

The relationship between freedom and agency in Figure 2.4 depicts the relationship between the concepts explained in this chapter and how the authors' approaches intertwine. The literature on capabilities and consciousness intersects and builds up a new approach perceived as the freedom of *consciousness-*

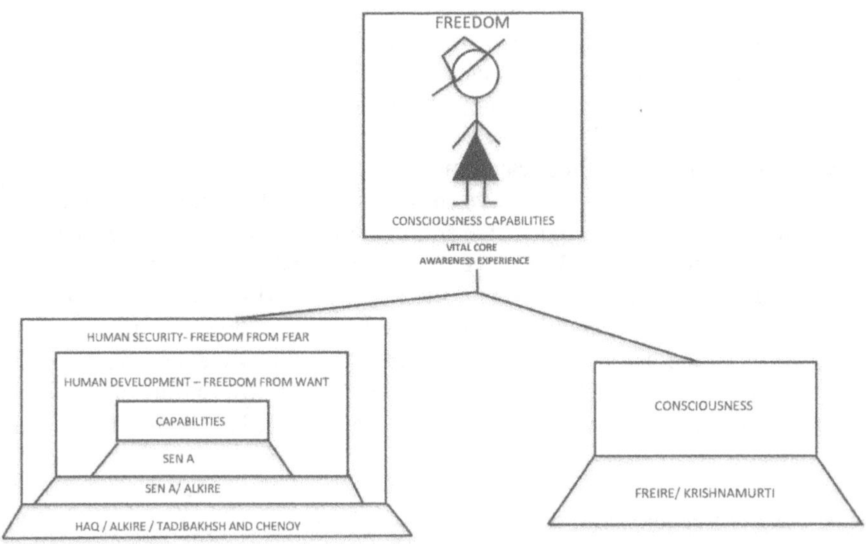

Figure 2.4 Building a multidisciplinary approach. ©Rodriguez Carreon, V. (2014).

capabilities. Consciousness-capabilities enable the freedom of the awareness of the experience impacting the *vital core.*

The analysis is multidisciplinary and cannot exhaust the existing literature, such as phenomenological consciousness. This study has relied mainly on the philosophy of Paulo Freire and Juddi Krishnamurti to explain the conscientisation in populations that experience oppression and the finding of fragmenting realities. Responding to the freedom from want is the human development approach based on the capability approach pioneered by the economist Amartya Sen. Human security, as the freedom from fear is based on human development and capabilities. The emerging theory of human security reflects the contribution of many authors from different fields. The literature on freedoms includes the ideas of economist Mahbub Ul Haq, who coined human security and was the founder of the UN Human Development Report. I also draw extensively on literature from Sabina Alkire, who supervised me during my visiting fellowship at the University of Oxford. Alkire, a member of the UN Commission on Human Security, emphasises the connection between the *vital core* and human security. The interdependence of the concepts in the literature is recognised to see the inner process of agency in its empowerment process of formation and transformation.

This chapter has set up the theoretical framework to analyse the three aspects—living in poverty, armed context, and seeking justice—for rural people's context to create a new understanding of their experiences. I have also emphasised the importance of perceiving the consequences beyond the physical and visible. At the depths of an agentive role of transformation lies certain motivations that are seeds in which flourishing depends on the capabilities to function in freedom and to exert agency. People become motivated because they value something to do or to be and have a reason for holding that value. The enabling of beings and doings those functionings with freedom forms a set of capabilities. The person can function with freedom in his or her agency based on this set of capabilities. Therefore, experiences within the capability set provide the person's well-being and the ability to decide the goals they value and have a reason to value to become the agent of their lives. Recognising this agency in decision-making is possible through the awareness of those capabilities.

Intersections between the literature on capabilities and awareness/consciousness was deliberately used throughout the following chapters to understand junctions among the person's agency. The concepts of agency and empowerment are increasingly debated within the development literature, primarily informed by Sen's approach to agency and capabilities. "This process of 'increasing-power' is conceived as the result of the interaction between two building blocks: agency and opportunity structure" (Samman & Santos, 2009, pp. 2–3). Haq's argument shows that agency could flourish in societies where the core is defined by human development and human security practices. Sen's and Haq's theories are based on capabilities because they are a people-centred approach. Capabilities are necessary for people who lack the freedom from want and freedom from fear. Awareness, on the other hand, is the concept intrinsically necessary for understanding the consciousness of the experience. Human agency and its rural human development perceived through the scope of *consciousness-capabilities* help us to examine our inner understanding of the *vital core* level of disconnection from its empowerment formation.

Notes

1 See more, Alkire and Deneulin (2009).
2 See more, Merleau-Ponty (2014) and Capra (2002).
3 This was noticed in my several fieldwork trips to Ayacucho from 2008 to 2013.
4 Framework for the working group session, a third in a series of United Nations Inter-agency Committee and OECD/DAC working party on gender equality workshops.
5 Testimony number 450075 given to the Truth and Reconciliation Commission while in investigation. Registered at the Ombudsman office.
6 Extra reading can be found in Chapter 3 Functionings and Capability. Amartya Sen, *Inequality Reexamined*, 1995.
7 I first heard this news on television in 2004, and then I read the article. I tracked the story to be able to document it here—see Beliakov, Dimitri (2004) *Tragedy in Beslam* in CBS NEWS. https://www.cbsnews.com/pictures/tragedy-in-beslan-2/
8 Alana had a brother two years old and, like many other older children, when the mother was asked to take the younger brother, the experience left her "dead inside". See: Murphy, Kim (2004) In rush of joy and guilt, a mother regains her child. September, 4. Web: Los Angeles Times: Russian hostage crisis. http://articles.latimes.com/2004/sep/04/world/fg-mother4; https://www.efootage.com/videos/34749/beslan-school-siege-survivors
9 In the case of late Mendoza de Ascarza, the disappearance was the injustice she pursued; the connection with her sense of agency kept her motivation for justice alive.
10 This perhaps can be related to the case of Christopher Gardner when he got out of poverty. The core of the motivation is in the love for his son, or in the willingness *to not give up* which is something that defines him, his core values written in his book: The pursuit of happyness.
11 See more, *Llaki* and *Ñakary*: Idioms of distress and suffering among the highland Quechua people in the Peruvian Andes by Pedersen, Keinzler and Gamarra (2010).
12 This quote is from Theidon, Kimberly (2004).
13 This is a summary of the Commission on Human Security report (2003).
14 This statement has been found in several NGO documents. The first one in the Strategic Framework for Peace Building–a development perspective by the Norwegian Ministry of Foreign Affairs: http://www.regjeringen.no/upload/UD/Vedlegg/Utvikling/peace-engelsk.pdf, and the second one, in "The role of education in Peacebuilding" is a literature review by UNICEF United Nations Children's Fund (2011), Presentation by: Stewart, Frances (n.d). *Conflict and Human* development.
15 The capabilities required to exercise consciousness are situated in the inner being, which should not be confused with work on perceptions of justice. The integration of justice will be explained in Chapter 5.
16 This is explained in detail in Chapter 5, "In the search for truth and justice".

References

Alexander, J. M. (2008). *Capabilities and social justice: The political philosophy of Amartya Sen and Martha Nussbaum*. Routledge.

Alkire, S. (2010). *Human development: Definitions, critiques, and related concepts*. May, 1. OPHI Working Paper No. 36. SSRN. http://ssrn.com/abstract=1815263 or http://dx.doi.org/10.2139/ssrn.1815263

Alkire, S. (2009). Amartya Sen. In P. Jan & S. Irene van (Eds.), *Handbook of economics and ethics* (pp. 484–492). Edward Elgar.
Alkire, S. (2003). *A conceptual framework for human security*. Centre for research in inequality, human security and ethnicity CRISE. University of Oxford.
Alkire, S., & Deneulin, S. (2009). The human development and capability approach. In S. Deneulin & L. Shahani (Eds.), *An introduction to the human development and capability approach: Freedom and agency* (pp. 22–48). Earthscan.
Beliakov, D. (2004, August 31). Tragedy in Beslam. *CBS NEWS*. https://www.cbsnews.com/pictures/tragedy-in-beslan-2/
Bohm, D., & Bohm, S. (1994). *Thought as a system*. Routledge.
Bohm, D. (1992). *Thought as a system*. David Bohm Seminars.
Capra, F. (2002). *The hidden connections. Integrating the biological, cognitive, and social dimensions of life into a science of sustainability*. Doubleday.
Commission on Human Security. (2003). *Human security now*.
Comisión de Entrega de la Comisión de la Verdad y Reconciliación. (2004). *Hatun willakuy* [The great story]: Versión abreviada del Informe Final de la Comisión de la Verdad y Reconciliación [Abbreviated Version of the Final Report of the Truth and Reconciliation Commission]. Navarrete.
Deneulin, S. (2009). Ideas related to human development. In S. Deneulin & L. Shahani (Eds.), *An introduction to the human development and capability approach: Freedom and agency* (pp. 49–70). Earthscan.
Deneulin, S. (2006). *The capability approach and the praxis of development*. Palgrave Macmillan.
Freire, P. (1973). *Education for critical consciousness*. Sheed and Ward.
Freire, P. (1972). *The pedagogy of the oppressed* (Trans. M. Bergman Ramos). Penguin Books.
Galtung, J. (1990). Cultural violence. *Journal of Peace Research*, *27*(3), 291–305. https://doi.org/10.1177/0022343390027003005
Gardner, C.(2006). *The pursuit of happyness*. Harper Collins.
Haq, M. (1995). *Reflections on human development*. Oxford University Press.
Herman, J. L. (1992). *Trauma and recovery*. Basic Books.
Lama, D. (2013). *Beyond religion: Ethics for a whole world*. Rider books.
Ibrahim, S., & Alkire, S. (2007). Agency and empowerment: A proposal for international comparable indicators. *Oxford Development Studies*, 35(4), 379–404.
Kieffer, C. H. (1984). Citizen empowerment: A developmental perspective. In Rappaport, J., Swift, C., & Hess, R. (Eds). *Studies in empowerment* (pp. 9–36). The Haworth Press.
Krishnamurti, J. (1982). *The network of thought*. Mirananda.
Krishnamurti, J. (1972). *The impossible question*. Krishnamurti Foundation.
Krishnamurti, J. (1969). *Freedom from the known*. Harper One.
Lederach, J. P. (2005). *The moral imagination: The art and soul of building peace*. Oxford University Press.
Merleau-Ponty, M. (2014). *Phenomenology of Perception*. Routledge.
Murphy, K. (2004, September 4). In rush of joy and guilt, a mother regains her child. *Los Angeles Times*. Russian Hostage Crisis. http://articles.latimes.com/2004/sep/04/world/fg-mother4
Navas, G., Mingorria, S., & Aguilar-González, B. (2018). Violence in environmental conflicts: The need for a multidimensional approach. *Sustainability Science*, *13*(3), 649–660. https://doi.org/10.1007/s11625-018-0551-8
Norwegian Ministry of Foreign Affairs. (2004). *Peacebuilding – A development perspective*. Utenriksdepartementet. http://www.regjeringen.no/upload/UD/Vedlegg/Utvikling/peace-engelsk.pdf

Nussbaum, M. C. (2011). *Creating capabilities the human development approach.* Belknap Press of Harvard University Press.

Nussbaum, M. C. (2010, February 2). *Examined life: Martha Nussbaum.* YouTube. https://www.youtube.com/watch?v=cbcGbflpFzI&list=PLvlPvvIcw-AHHSeW1P53AvxpotblD-7VD

Pedersen, D., Kienzier, H., & Gamarra, J. (2010). Llaki and ñakary: Idioms of distress and suffering among the highland quechua in the Peruvian Andes, *Culture, Medicine and Psychiatry*, *34*, 279–300. http://www.ncbi.nlm.nih.gov/pubmed/20405314

Petri, H. L. (1986). *Motivation: Theory and research.* Wadsworth Publishing Company.

Porter, E. J. (2007). *Peace building: Women in international perspective.* Routledge.

Rodriguez Carreon, V. (2014). *Empowerment formation: women's agency for participation in decision making within the poverty and conflict context, case of rural Peru* [Unpublished Doctoral thesis]. The University of Sydney.

Samman, E., & Santos, M. E. (2009). *Agency and empowerment: A review of concepts, indicators and empirical evidence.* Oxford Poverty and Human Development Initiative OPHI. University of Oxford. http://www.ophi.org.uk/wp-content/uploads/OPHI-RP-10a.pdf

SASC. (2014, June 4). *Post it notes analogy for effects of trauma on memory.* Dr. Rebecca Campbell discussing how trauma affects our ability to store and retrieve memories. From presentation - the neurobiology of sexual assault. YouTube. https://www.youtube.com/watch?v=AnRpoGSOkZs

Schaub, F. (2012). *The fear and anxiety solution.* Sounds True.

Sen, A. (1999). *Development as freedom.* Oxford University Press.

Sen, A. (1995). *Inequality Reexamined.* Oxford University Press.

Stewart, F. (2002). From reaction to conflict prevention: Opportunities for the UN system. In F. O. Hampson & D. Malone (Eds). *Horizontals inequalities as a source of conflict.* Lynne Rienner Publishers.

Stewart, F. (n.d.). *Conflict and human development* [PowerPoint slides]. UN Development Programme, Presentation. https://www.powershow.com/view/42b3-NTQ3O/Conflict_and_Human_Development_powerpoint_ppt_presentation

Strandberg, N. (2001, November 19). Empowerment of women as transformative strategy for poverty eradication. In *United Nations, Division for the Advancement of Women (DAW).* EGM/POV/2001/EP.6.

Tadjbakhsh, S., & Chenoy, A. M. (2009). *Human security: Concepts and implications.* Routledge.

Theidon, K. (2004). *Entre prójimos: El conflicto armado interno y la política de reconciliación en El Perú.* Instituto de Estudios Peruanos.

Hanh, T. N. (2003). *Ending violence in yourself, your family, your community, and the world: Creating peace.* Free Press.

UN Office of the Special Adviser on Gender Issues and Advancement of Women Department of Economic and Social Affairs. (1999, December 7–8). *Women's empowerment in the context of human security.* ESCAP. https://www.un.org/womenwatch/ianwge/collaboration/Rep1999_WE.PDF

UNDP United Nations Development Programme. (1994). *Human development report 1994.* Oxford University Press. https://hdr.undp.org/system/files/documents/hdr1994encompletenostatspdf.pdf

UNDP United Nations Development Programme. (2005). *Human development report 2005.* ISBN 0-19-530511-6. https://hdr.undp.org/system/files/documents/human-development-report-2005-english.human-development-report-2005-english

UNDP United Nations Development Programme. (2021/2022). Uncertain times, unsettled lives: Shaping our future in a transforming world. *Human development report 2021/2022.* https://hdr.undp.org/system/files/documents/global-report-document/hdr2021-22pdf_1.pdf

UNICEF United Nations Children's Fund. (2011). *The role of education in peace building: Literature review.* http://www.unicef.org/education/files/EEPCT_Peacebuilding_LiteratureReview.pdf

Varela, F., Thompson, E., & Roch, E. (1993). *The embodied mind. Cognitive science and human experience.* The MIT Press.

Watters, E. (2010). *Crazy like us: The globalization of the American psyche.* Free Press.

Part II

Context

Dynamics of Human Insecurity

3 Poverty and its forms of oppression

This chapter explores the constructed identities attributed throughout the history of oppression to the women from the highlands who live in underdeveloped infrastructure and without access to government institutions where their livelihoods are associated with being poor. Also, I review the core values that are missed in their identities. Because of racial oppression, Quechua women's identities experienced dynamics of disempowerment, and when they interacted outside of their rural context, their identities (as how they see themselves) received negative connotations, what I have called "disempowered connotation". Racial oppression exists in the interaction with a dominant norm and forms of powerlessness (Young, 2011). In Peru, identities connoted as marginalised and its structured forms of oppression constituted *being* women, *being* from Oral cultures (Quechua, Aymara, and others), and *being* from the mountains or jungle rural ecosystems with access to limited economic resources. The structured unequal value given to the rural women's identities is the result of the negative connotations attached to their gender, ethnicity, and the constructed label of "class" or place to access an economic network of resources—the connotation of these categorisations is part of the problem I found in defining societies in a narrative that depends on the hierarchisation of identities. In the stories shared in this book, the experiences of disempowerment living in Andean women's bodies are reproduced mainly within a context where *being* and *doings* as a Quechua are connoted as powerless. The disempowerment dynamics stem from beliefs based on oppression manifested in educational, cultural, and political dimensions.

Negative connotations attributed to physical bodies and these bodies' ways of expressing their being perpetrate an Othering and simultaneously attribute a hierarchical value to identity and its access to power. The Other, the peasant, is not seen by the culture of domination as being entitled to power. Or is it the Other who is in power over to create the narrative of the Othering? Oppression constrains peasants' agency because the perceptions of their identities have been gradually disempowered over the centuries. While the actions to oppress came from as far back as colonial times through policies, they are perpetuated until today through unconscious ways of perception encoded on the individual and social body. Living in poverty and the consequence of its oppression obstructs the challenge of the empowerment formation and, therefore, the development of the rural people's agency.

DOI: 10.4324/9781003450955-5

For Iris Young "oppression has often been perpetrated by a conceptualization of group difference in terms of unalterable essential natures that determine what group members deserve or are capable of, and that exclude groups so entirely from one another that they have no similarities of overlapping attributes" (Young, 2011, p. 47). The identity of women from highlands has been categorised and conceptualised with determined qualities as women, Quechua and living in poverty. Each of those characteristics are ways of being with individual and collective attributes grouped and encoded hierarchically into the consciousness of people. The historical narrative depended on constructing and connoting those identities as oppressive. The question is, by whom? Who gets to define identities during this period of oppression? The dynamic to portray certain identities with a negative connotation resulted in women's agency remaining in a passive state and blocked from the natural development of their agency.

Feeling identified as a woman, or Quechua, or living with limited resources does not resemble oppression. The problem is that these identities in determined contexts were deprived and oppressed systemically and are not freed from the identity being associated with access to power. Acknowledging the identity of women, or Quechua, with their values is freedom. The freedom of people from structural oppression is necessary to enhance the empowerment process of those identities defined and associated as oppressive. Paulo Freire proposed that:

> The situation of oppression is a dehumanized and dehumanizing totality affecting both the oppressors and those whom they oppress, it is the latter who must, from their stifled humanity, wage for both the struggle for fuller humanity; the oppressor, who is himself dehumanized because he dehumanizes others, is unable to lead this struggle.
>
> (Freire, 1972, p. 24)

Freire's analysis calls for a freedom from oppression in which the oppressed must be the protagonist of their own story, and of their own freedom. At the same time, he recognised that both oppressed and oppressors are trapped in a "duality" of beliefs; they want freedom and at the same time they fear freedom from beliefs with which they became familiar during the time lived under oppression. I argued that the oppressor is responsible for their own struggle. This inner conflict alters the quality of connection with the *vital core*, with the fears of the oppressed driving the transformation and the oppressor fearing to be controlled by those they oppressed. The detachment from those fears, however, will enhance the freedom of their agency, for both oppressed and oppressor.

Attachment to fear is a fragmented thought, which, in Freire's analysis, is as being "immersed" and "resigned" to the "structure of domination" (Freire, 1972, p. 24); the context where rural people, for example, were stuck since the colonial period. Identities of people living in structural violence, as it is poverty, encode inner qualities of disempowerment that submit their identities and reduce their agency's image to the outer world. The unequal treatment of women in Third World cultures under colonialism led to female poverty (G. Sen & Grown, 1988, p. 30). Women—in particular peasants—were dehumanised within Peru's oppressive system. Since colonialism, oppression has embedded the idea of obedience from the peasantry to patriarchal power. It justified aggressiveness against different people in comparison to Hispanic tendencies, whose identities enjoyed a powered connotation. Structural and hierarchical marginalisation emerged from the domination of the colonial mindset and became the root cause of the disempowerment of Indigenous, peasants, and/or rural people.

Dynamics of oppression conditioned the perception of people's experiences associating and connoting to a negative sense of Self. It is important to note that the peasantry may enjoy their gender and ethnic identities when they are in familiar contexts (see more, Thorp & Paredes, 2010; Webb, 2011). For example, rural women's perceptions of their own identities play a role in creating an empowered identity when interacting within the same community. This is achieved by forming associations where they exercise agency by advocating common concerns, such as the need to search for those who survived the conflict. Nevertheless, when in their peasant roles interacting within different spaces, such as in the urban environment for example, then disempowerment dynamics may be experienced through oppressive attitudes exercised against their customs, language and race. The perception is not in freedom. Instead, it is conditioned to the perception of the dominating view and consequently, depending on time and space in a determinate context, the identities are connoted hierarchically and, with that, associated with power.

Highlighted here is the importance of acknowledging agency to enhance women's empowerment to prevent their victimisation. A positive valuation of women's identities is possible by valuing the identities of those living in poverty through acknowledging their agency exercised within their communities.

The analysis focuses on how the architecture of oppression, in context and content, disempowered rural human agency in Peruvian history. For example, marginalisation and its connotation for the disempowered identities structured oppression from depriving access to the system's decision-making, lack of opportunities in education, cultural violence and political restrictions, to name a few. Structural manifestations focus on the barriers found by women who challenged their marginalised identities. Here the concept of poverty is interpreted as the lack of human development that reduces the chances to develop agency. Empowerment is needed in rural human development to reinforce freedom from "the oppressor whose consciousness they have internalized" (Freire, 1973, p. 24). Structural oppression is embedded in systems and represses the capabilities of being free through reproducing power within the system's stratification. It is called power structures to the State institutions loyal to the systemic doing that permeates the structure as impersonal. Therefore, rural agency for decision-making in this chapter is mainly passive on the outside, and yet active within. This chapter examined how women's experiences built their communities and how they challenged the barriers of poverty's disempowering effects when oppression is still happening, and armed conflict unfolds.

Rurality in Peru: marginalised identities, for who?

> The tragedy of Indian success stemmed ultimately from the way it secured the participation of a defeated people in its own oppression. The colonial regime rewarded Indians whose advantages, skills, or luck enabled them to

> adopt Hispanic forms of accumulation, and punished those whose identification with the peasantry was too strong or aggressive. The political implications were profound. The lure of success and the threat of loss recruited Indian allies to the colonial power structure, discouraged overt challenges that invited repression, and fragmented the internal unity of Andean society. The economic implications, too, were far-reaching. Success stimulated class differentiation within the "republic of Indians," dividing it into the rich and more acculturated on the one side, the poor and less acculturated on the other.
>
> (Stern, 2005, p. 147)

The statement by Stern about colonialism captures the historically structured oppression of the peasant's identity over centuries. Thus, the peasantry's identity evolve for themselves in contact with the land, yet they were connoted by the regime as disempowered. Their inner qualities of connection to the land and their technology in working with it reflected the knowledge and power of their Oral culture. However, in a recent paper I wrote how "colonialism alienated oral cultures and bestowed a sense of superiority to those able to read and write words, whilst denouncing other ways of being and other forms of symbolizing language as inferior ways to sense and to communicate. Indigenous cultures across the world were repressed in their ways of expression" (Rodriguez Carreon & Vozniak, 2021, p. 38). The culture of ignoring Oral wisdom, and associating power and education with reading and writing gave the colonials a corrupt sense of superiority. Yet, systematically implemented through colonial policies first, coercive measures towards peasants promoted the conditions that both marginalised and segregated the identity of the poor as someone without knowledge. Systemic measures were implemented, for example, through the imposition of taxes; the amounts were higher than the salaries paid by colonial authorities, and the introduction of policies that disabled opportunities for access to land by peasants. Moreover, what enabled a colonial mindset was the imposition of the "need for goods" that were either owned by Hispanics or connoted Hispanic trends and the provision of "credits" through resources that created debts impossible for the peasants to pay (Stern, 2005). The peasants were oppressed and learnt in this alienation what Paulo Freire calls the duality, *to be* is *to be like*, and *to be like* is *to be like the oppressor* (Freire, 1973, p. 25). Freire (1973) noticed that as long as this division is made, it is impossible to free the Self.

Followers of Andean trends were under surveillance and receiving constant threats that disempowered to be recognised for their identity as peasants. Direct and subtle divisions were between those who maintained their Andean identity and those who felt coerced to be alienated by Spanish trends. In addition to this coercion under colonialism, exploitation was assumed as the only way to succeed in escaping from poverty. Profiteering was possible through the creation of private business with Hispanic tendencies. Coercive measures included hiring peasants for extremely low wages, and other actions exercised by those in the position of authority in the institutions. To avoid the exploitation, the peasant felt obliged to agree with deals that put them at a disadvantage. They had to maintain economic, social and political Hispanic cultural appropriateness; otherwise, repression was used to undermine their Andean cultural way of living. The power to succeed therefore was left to those who exerted dominance over the peasants within the different dimensions of power (Stern, 2005). The damage from oppression arose from the continuing use of mechanisms where the perception of the Quechua was maintained as disempowered. Those who were in positions of domination and colonial way of living undermined and disempowered the peasants, Indigenous and rural identity.

The Andean people had in common customs, race, appearance, and language. Those communal characteristics came under threat during colonialism and created divisions among the Peruvians. Indians, as Stern (2005) refers to them, who were willing to adopt Hispanic colonial forms of living, were closer to becoming "respectable" citizens back in colonial times. The metamorphosis caused by colonial dominance left its legacy of disempowerment for those that wanted to survive with the Quechua identity. Since then, the identities of the population have been subjected to diversification, without altering the condemned perception of those identities' "inherited degradation" (Stern, 2005, pp. 124–148).

However, in radical different visions, both the Inca and Colonial systems have wrought a gender division among the Andean people, and since then women have been deemed inferior (Silverblatt, 1987). Felipe Guaman Poma Ayala was a native Peruvian Andean writer who argued that his main aim in writing was to "retell" the history of Colonialism. He wanted to show the difference to those discourses with Hispanic tendencies (Adorno, 1986). "The social relations into which Andean women and men were born highlighted gender as a frame to organise life … Women and men acted in, grasped, and interpreted the world around them as if it were divided into two interdependent spheres of gender" (Silverblatt, 1987, p. 20). Women and men respected their conceptualisation of the universe and their relationship to it. Each gender has its own God, the *Inti* (Sun) for the men and the *Pacha* (World, Cosmos, Universe) for the women, and men and women have had autonomy in their respective genders. Yet, Irene Silverblatt adds, when the Incas built their Andean empire, they did so due to gender divisions, partly through the "rendering of virgins to the Inca's service" (Silverblatt, 1987, p. 85); gender differences achieved a major transformation in the lives of the Andean peasants. Andean women with "physical perfection" and "social rank" were the *acllas* chosen by the Incas to be the Sun's wife and, above all, they were virgins compared by the colonials to "Rome's vestal virgins or with the nuns of the Roman Catholic church" (Silverblatt, 1987). Guaman Poma criticised the colonisers, and pointed out that Andeans, "suffering, degraded, and dehumanised", were described as being without voice and hope (Adorno, 1986, p. 129). With colonialism, oppression transformed gender divisions with the introduction of religion. Colonisers demanded several measures against women. Andean women did not pay taxes during the Inca period. "Women had honour", but during colonialim "the priests want[ed] money, money, and more money, so they force[d] women to serve them as spinners, weavers, cooks, breadwinners" (Silverblatt, 1987, pp. 125–131). Women were gradually defiled by colonial measures and categorised as inferior, as they became obligated to pay taxes. Silverblatt (1987) stated that the colonial regime was "so devastating to indigenous society and so fundamentally inhumane" (p. 126).

The hierarchies structured were different from the categories that existed before the colonial times. Discrimination against ethnicity, access to financial power (class) and gender identities brought degradation to the connotation of the peasant's identity. For example, the patriarchy segregated gender, situating women as an oppressive identity. The hierarchical fragmentation of identities in the patriarchal system triple marginalised illiterate, rural, and female. These historical facts explain why and how societal structures of oppression were formed through institutionalisation as a relational method to control communication with the peasantry. The last decade when my fieldwork ended, rural women in the highlands were still excluded and marginalised from the main basic services such as water and electricity, or access to school and health services. The survival agency that some identities faced in overcoming the obstacles from stigmatised oppression and marginalisation challenged the inner architecture.

The identities and their disempowered connotation

Often, rural people who hold back in their agency due to structural oppression, face situations of disempowerment in the cities, where their identities are seen through a hierarchical perception and deemed inferior by the system that withholds the dominion of their identities' narrative. For any identity, experiencing oppression means being exposed to mistreatment that is held on scarcity and undermines the evolvement of human agency. Several times while living in Peru, I witnessed and confronted the plausibility of unequal identities perceived on economic conditions, which were structured across all systems, social, cultural, and political. For instance, over two decades ago, when a male guard was working on an affluent business building, two women attempted to enter the carpark. Their attempt was frustrated by the guard as they were not the car's owners. Unsatisfied with the lack of permission, one of them started a one-way argument with the guard, and called him *cholo* and ignorant. Then, the guard moved his head down. In Lima's culture of domination, his identity through the filters of the triple disadvantage connotations was seen as powerless. The two women's appearances associated with identity hierarchies claimed entitlement to power over other identities. The "duality" of freedom here results from two identities that are historically constructed. The guard on duty exercising his responsibilities was disempowered when he was told *cholo;* the word was connoted as an insult. *Cholo* is cultural contextual and has several meanings related to low social economic power, particular ethnic features, gestures, language, and educational references. This depends on where it is being used, which is important to note the context of domination. Oppressing his identity based on attributes associated with disempowerment features situated the oppressor's cultural condition in the urban context with over powering. At the same time, the other felt entitled based on their beliefs supported over the history of oppression from systemic to a collective unconscious agreement constructed to favour some identities over others. Several hidden dimensions transmit messages in everyday interactions of disempowerment; for example, how culturally is power associated with entitlement rather than weakness? Experiences in the urban day-to-day interactions to name someone "ignorant" is an insult and it is connoted as unknowledgeable, and having knowledge is connoted as positive—knowledgeable meant being Spanish literate, therefore, educated. Yet, the reaction, at times, when listening *cholo* (masculine)/*chola* (feminine) remarks is to respond with the same statement. In doing it, I argue that there is an unconscious consensus that the term *cholo*/*chola* is connoted powerless and, therefore, it is sealed in internalising with a negative association. Unconsciously, power is associated with positive remarks. However, the identity of *cholo* may enjoy a positive connotation within a different environment. There is a narrative constructed around the identity of those in vulnerable environments that needs to be connected to what those that identify with the term are what they value and have a reason to value.

Here, I want to *pause* to reflect on alienation: How can a different narrative rebuild the collective Othering that had internalised concepts of weak = ignorant = unknowledgeable in Spanish (or former colonial language) identities as powerless? If person A says you are "ignorant" out of fear, the person A's existential fear is being "ignorant". Alienation to an identity with Spanish trends was a matter of life and death in colonial times. Othering someone by naming them "ignorant" was assimilated as a protection from being associated with an "unknowledgeable" culture. Alienating from anything unrelated to the narrative of domination, culturally, meant to embody weakness. Today, collective Othering, caused by associating identities with negative connotations, is a sign of weakness in healthy society seekers. Paradoxically, to be in the unknown is the inner quality needed for human development.

When public servants exercising functions within the system produce and/ or reproduce disempowering interactions with the rural citizens, they fragment and exploit the decision-making power perception. Young's work understands the impact of oppression as structural problem with social, political and economic impact on the agency of those where stereotypes are embedded in exclusion (Young, 2011).

The perception of an identity and its "value" are, in particular, linked to gender, ethnicity and access to financial resources. When the triple disadvantage endures the mistreatments of oppression, as Young writes, the oppressor does not necessarily exercise his or her actions consciously and the oppressed may not know that such action was exercised unconsciously. Also, "while structural oppression involves relations among groups, these relations do not always fit the paradigm of conscious and intentional oppression of one group by another" (Young, 2011, p. 41). However, structural oppression in Peru is exercised on a day-to-day basis, and its functioning is based on unconscious social agreements in various spheres of life.

Oppression working as disempowering identities in relation to gender, ethnicity and access to socio-economic power became the ill inner noise to undermine agency. The obstructed parking guard's agency and his reaction can be understood by perceiving it through the following statement from Young: "in the most general sense, all oppressed people suffer some inhibition of their ability to develop and exercise their capacities and express their needs, thoughts and feelings" (Young, 2011, p. 40). It can be analysed that instead of saying that when the guard moved his head down, he manifested his internalisation that negatively connoted an identity associated with a disempowered inner condition; it can be said the expression is an existential symptom, whereas suppression is happening, he is protecting his inner condition—both interpretations are in *survival agency*, and also the oppressors are on survival because their power over is a symptom of fear *to be like*. It is important to note that, all people with similar characteristics do not feel and react the same way (Young, 2011). This depends on how people see and perceive their own identity, as they have a reason to see themselves in many ways (A. Sen, 2007, p. 15). Also, it depends on the embodied and embedded experiences that conditioned entire cultures since colonialism, as it is the case of Peru, and on their awareness level of such conditioning. For example, the capability to be perceived as peasants in connection to the land with empowered connoted identities was obstructed by the centuries of structural marginalisation across spheres.

This chapter sets the context of structural marginalisation within the culture of domination towards identities with disempowered connotations analysed in three forms of identity: gender, ethnicity, and access to social and financial network of power (class). Also, I noted the effects of oppression that depend upon how much internal engagement occurs with disempowered identities at different levels. Suppose common identities are oppressed within a group, for example, being a woman, when women are harmed, and their attachment to the identity is strong; in that case, the harm of oppression affects the *vital core* of their human dignity. Being Quechua is NOT a disempowered identity. Within the cultural, social, and/or political dimension of life identification, peasants in their communities enjoy the freedom to relate to their identities with empowerment. But, during times of armed conflict in the urban areas, to be Quechuan and a woman were identities subjected to marginalisation.

How identities and their disempowerment connotation play a role in the oppressive context is analysed here in relationships. Identities are to be seen as interdependent in their context (Burke & Stets, 2009, pp. 3–9), and in relation to something or someone (Young, 2011, pp. 45). The possibilities of each identity rely on individuality, and within the situation that predestines potential capabilities (A. Sen, 2007, p. 5). Then, marginalisation bestowed among identities can challenge or hinder depending on the contextual dimension in which it is exercised. Young affirms that joining an association affects a person's life although the group affinity does not determine a person's identity (Young, 2011, p. 46). However, colonialism went beyond expansion. Colonialism is a mindset that aims to restructure intersecting identities hierarchically. It included collective identities, for example, with a determined institutionalised religion. A colonial system meant the imposition of internalising a hierarchical perception over another person and/or culture, mutating from a scarce inner source of dominium while extracting, expropriating, and expanding their knowledge of the Other.

Gender

Gender is an identity category systematically constructed over centuries. In Peru, the division among the Andean people under the Incas rendered "gender distinctions into approximate images of the class relations they [the Incas] imposed. Gender hierarchy and class formation were merged, perhaps irretrievably, as the Incas forged their Andean empire" (Silverblatt, 1987, p. 108). With colonialism, gender divisions changed; for example, being privileged depended on women's colonial connections. Colonialism "exploited" the Andean people, and the Spanish system "imposed special burden on women". As stated by Doña Paula Mama Guaco Ñusta (ADC: ACC, top. 9, Leg 5), "Certain Indian women … dishonoured me with insulting words, ignoring the fact that I am the honourable wife of a Spaniard, a women of quality and nobility" (Silverblatt, 1987, pp. 109–111). To be a woman was not the only identity to bring a division, yet to be a woman of class. Economic hierarchies were labelled, and class as "status" was for those associated with Spanish links.

For rural people gender was a strong identifier. Thorp and Paredes (2010) did a study in identity in rural Peru and found that Ayacuchanas (women from Ayacucho) saw themselves first as women (Thorp & Paredes, 2010, p. 16). Burke and Stets (2009) write about how humans identify themselves. Identification is based on perception and gender, as identity is stored in memory (pp. 63–64). Andean women and their identity connotation with class were also challenged. Over time, diverse associations and symbols are compared and conceptualised, entangling socially a hierarchical view of the Other identity.

To perpetrate violence to the identity of gender, it had to be connoted as inferior through the imposition of *machismo*, a patriarchal behaviour coerced most by men. For Gerda Lerner this male superiority sensed from biological factors have its "primitive" narrative in the Stone Age. Superiority in strength, aggressiveness and faster running were attributes prized for the hunter provider of food and therefore received value and honour over women's motherhood and nurturance (Lerner, 1986, p. 17). At the same time, nurturance did not enjoy the same perception of value, resulting in the primary differentiation in perception that connoted the identity of the woman's body and the feminine characteristics as inferior. The colonial mindset resulted in associating identities of power with ways of being that enjoyed positive connotation. Consequently, when violence was perpetrated by those perceived as possessing superior physical strength and aggressiveness, it transformed the culture of gender. In rural Ayacucho, it was women's partners, often under the influence of alcohol, who largely perpetrated domestic violence. Domestic violence towards women is an example of disempowerment in their agency relationship. According to William Mitchell (2006), beating the wife was not seen as natural in the communities. It was viewed as a consequence of consuming alcohol and the legacy of poverty and inequality (2006, p. 45).

In relation to the perception of identities, I ask, in war, which of the many identities is connoted as being powerful in the perpetrator? In Theidon's work in rural Peru, a woman crying said, "*mamay, imapapaqtaq warmita wacharihuaraki*" (Mami, why did you give birth to a girl?). The girl said it to her mother after being raped by the military forces during the conflict (Theidon, 2004, p. 125; Theidon, 2013, p. 127). The identity of being a woman in the highlands was felt very strongly. However, in poverty and conflict, many victims in rural Ayacucho, rejected their identity as a woman. Theidon brings the example where the woman does not want to be or to be connected with being a woman. The woman believes that such abuse would not have happened if the identity (gender) had been removed. This inward feeling is a passive state in the outer. There is a split on the level of attachment to what represented being a woman. At the same time, it depends on the strength of the affiliation or core of values as to how their gender was compromised and how it affected their *vital core* through their experience.

Some perceptions of gender are deeply rooted. For example, many peasants' children, whose conditions of poverty limited their future in terms of economic benefits, seek prosperity in urban areas. Rural women living in poverty who move to the urban areas take employment in housekeeping, cleaning, and other service jobs. In many cases, access to this "developed" environment, urban life, was with mistreatment from their employers. Urban structural oppression exposed them to psychological and, at times, physical workplace violence. Violence towards rural women, then, leaves them with lack of access to justice for their ownership of their own bodies. Trying to access justice was so disempowering that it solidified on them the consciousness of being in an environment of oppression. As explained, oppression requires a sense of dominance over the one who was felt by the oppressor as unequal and inferior (Young, 2011, pp. 58–61). Sensing this disempowered connotation over and over again constructs the structural wall of unequal values to gender identity.

Women in remote rural communities experienced the horror of conflict at an abhorrent and unspeakable level of human hatred. A woman who was three months pregnant when her husband was killed, was taken by the *terrucos* to work as a slave, but managed to escape, only to be taken by the military and, after all her suffering, asked for things for her children and her husband's body but nothing for herself (Henriquez Ayin, 2006, p. 34). Her statement shows her capacity building on her resilience based on her survival agency. Rural women who experienced repression, aggression and other injustices constantly during conflict attributed their way of living in poverty to be the reason for these unjustifiable experiences.

The soldiers in the military forces failed to identify the humanity of those identified of being a woman and raped them, referring them as things. The violence exposed the military forces and the Shining Path's inhuman condition that enabled them to perpetrate the most grotesque acts. Women were oppressed through violence, disempowering them to the level of making them doubt their own human condition as women.

Ethnicity

The identity connoted as disempowered has evolved since colonial times. When the Spanish segregated and oppressed different Indigenous cultures to be their servants and built hierarchies between them and others in line with the shades of skin colours intertwined with ways of beings and doings, simultaneously, social meaning was constructed and embodied according to the colonial frame. Indigenous people's physical characteristics were associated with a disempowered meaning. Those who did Othering in others' ethnicity during the war identified first their physical impressions, skin colour, and exposure to work under the sun and land. Hence, the oppressors who carried out marginalisation made categorisations inwardly. On the other hand, their racial appearance could be recognised as indicating someone hard-working and having a different connotation among the fellows in the same community. The question is, how do we see people the way they see themselves? The disempowered ethnic association of the Quechua's identity by those who were in domination of the Peruvian system subjected them to humiliation. Centuries of oppressed ethnic identity combined with the deprivations from living in poverty put the Andean population into a disempowered connotation. Rural people have not received an acknowledgment as empowered identities for centuries.

Thorp and Paredes (2010) conducted a study on identity and found that ethnicity is a complex topic in rural Peru. They acknowledged that "given [the] discrimination and a history of oppression, how people speak about their identity is never straightforward". *Ayacuchanas* defined themselves in terms of their gender, occupation (their doings), and ethnicity. Before the displacement originated due to conflict, these three identities were generally combined into one. In Thorp and Paredes study people from the mountains defined themselves as *andino* (from the Andes), *serrano* (from the Sierra) and indigenous (Thorp & Paredes, 2010, p. 16). *Andino* and *serrano* are terms that allude to being people from the mountains or the highlands. The three characteristics are differentiated, but they may not have the same categories within a different context. Place of birth, mother tongue, and racial origin categories, or as suggested by Sen, "classifications", are defined in terms of identities (A. Sen, 2007), and when combined into one particular identity, is ethnicity.

Social, cultural, and political dimensions of identity vary depending on time and place. Stratifications among Quechua people may also exist within the urban communities and are complex. For example, in regional Peru, Andean women interacted with other women who might live in the city province, Huamanga, the capital of Ayacucho. The women from Huamanga may consider themselves more urban, so the perception of their identity may not be the same as the rural women coming from the provinces in the Ayacucho mountains. Either from a small community or from the main town, at every level, there exist different perceptions of identity. Quechua native speakers living in urban communities may not feel rural or with the identities perceived by the rural people from isolated communities. Also, rural women may perceive themselves with other identities not seen by others who do not share the common context, like in the capital. For example, knowing Quechua (language) may be a reason to feel unsafe from experiencing racism in Lima (the capital of Peru), but not in Ayacucho (Thorp & Paredes, 2010). Othering gradually built its inner architecture according to time and space and the level of attachment to identities.

What is the people's freedom to maintain their identities with value in the eyes of others? Amartya Sen (2007) argument about the freedom to value is relevant here. During times of conflict, under "bellicose identity" we may have to challenge our other senses of identity. Affiliations with one particular identity–bellicose identity–risk the influence by others on exploiting such categorisation under extreme conditions (pp. 4–15) of being the person's sole identity. The ethnic categorisation of being Quechua in Lima is subject to anarchical aggression and, at the same time, the moral regression of those who share the reason to value their ethnicity.

The Quechua ethnicity was endangered during the Peruvian conflict. Enabling a narrative to perpetuate violence against them, targeting and identifying Quechua ethnic features became a method to degrade and dehumanise them.

The Truth and Reconciliation Commission (TRC) exposed how the media television gave attention to the conflict when tragedy reached the urban areas, and the victims were not the *campesinos Quechua hablantes* (Quechua peasants speakers) but the Hispanic population (Comisión de Entrega de la Comisión de la Verdad y Reconciliación [CVR], 2004, p. 21). I am not saying here that the urban population who also experienced violence should not be shown in the media, but to expose the identity value of all victims-survivors and their tragedies. Quechua, Aymara and other oral cutlures did not enjoy full citizenship, and their identities during conflict were excluded and marginalised when they interacted in other environments, where the legacy of oppression has been overshadowing them since colonial times.

Class (access to resources with power over economic network)

Colonialism brought the concept of class, a structure to categorise hierarchical beings and doings of some identities to benefit others in order to get access to economic power resources. Racism, direct or subtle, relies on class to suppress one humanity for another and subvert other ways of power. In Peru, Andean women are marginalised when they move to metropolitan areas due to their social and cultural categorisation internalised by others and their resistance to feeling alienated from Hispanic trends. This alienation occurs when there is a lack of awareness of "the shift" or "the transfer" of the person and her or his personality (Rotenstreich, 1989, p. 77, 45). For example, alienation would be "the transfer" of sensing being a Indigenous, peasant and/or rural to not being one. It is "the shift" which many felt coerced to do it due to the oppressive connotations of their identities in class and culture. During colonial times, the change to Spanish trends was part of the policy. After independence, it became part of the acculturation; today, it is unawareness of the unconscious embodiment, individual and collective.

Since colonialism 500 years ago, the system of domination had discriminatory remarks for the Indigenous cultures, for their Andean clothes, their orality, weaving technology, as well as their resistance to the city's customs, and, in general, their limited Spanish ways of being. At a national level, they were deprived of interaction with power structures. Most of these women did not have interactions with the positively connoted called urban development infrastructure and services. However, a person from the city who has never been in the mountains would have limited knowledges about the culture and geography, yet they are not connoted as disempowered. Even so, as urban people have access to books, online services, and unlimited information, if they attempt to interact, they have a better opportunity to access the knowledge of what to expect from rural areas. When peasants, rural and/or Indigenous people are out of their environment in the highlands, they are stigmatised for their class, connoted as disempowered by their appearance, and categorised as living in poverty.

Class is a hierarchical structural constructed identity, where the underclass is seen as disempowered. Young pointed out that people who were born at the bottom are more likely to be the oppressed class (Young, 2011). Paul Collier (2008) coined the phrase 'the bottom billion' to describe the world's populations trapped in poverty. People living in poverty are marginalised for their status and the meaning carried within it. They are not allowed autonomous interaction—freedom from oppression—at the urban level. Urbanity beyond material infrastructure involves mechanisms and strategies designed and institutionalised to deal with language, customs, and other cultural constructions. I argue that learning urban social capital is possible if obstacles such as exclusion and marginalisation do not judge and undermine the core identity of the rural human development and agency.

Thus, to be poor is to be vulnerable and targeted as an oppressed class in the urban context. The level of oppression in class depends on the infliction, whether it is chronic, ongoing, or seen and acknowledged. Porter writes that even though marginalised people experience oppression in different ways, they suffer common obstacles for the full realisation of their "self-development" (Porter, 2007, p. 70). It is still possible to feel empowered to pursue freedom from class deprivations as long as the agency's *vital core* is not affected.

In conflict times the consequences were greater for those at the bottom level of resources with power over economic network. Amartya Sen has written, "the poorest members of any community are the easiest to kill" (A. Sen, 2007, p. 173). When conflict occurred then, women experienced violence at unimaginable levels of pain and suffering, which reminded them of their marginalised class. Dehumanisation in war reminded people living in poverty of their "assigned" status, given by the dominant cutlure. The major tragedy for the TRC was to discover that most of those who died during the conflict were the invisible Peruvians, the Quechua peasants. Mrs. Sabina Valencia TRC testimony in Quechua and Spanish said:

> Senores chaymi nuqa munani kachun respeto, kachunya manchakuy, masque imayrikulla kaptiykupas, wakcha pobri kaptiykupas, campesino totalmente nuqanaykuchu kaniku, huk real llapas killapi ganaq, mana ni pipas kanikuchu. Senores, chayta ya justiciyata manakuykiku [Señores, por eso yo quiero que haya respeto. Que haya pues temor de Dios, aunque solo seamos muy humildes. Aunque seamos huérfanos y pobres. Campesino puro podemos ser; que ganamos solo un real por mes y, aunque no seamos nadie, señores, esta es la justicia que le pedimos].
>
> (CVR, 2004, p. 27)

In English:

> Gentlemen, that's why I want to be respected. So let there be fear of God, even if we are only very humble. Even if we are orphans and poor. Pure peasant we can be; that we earn only one real per month and, although we are nobody, gentlemen, this [respect] is the justice that we ask of you.

Mrs. Sabina is saying that even though they are living in poverty, they are human beings. The dynamics of disempowerment for those living at the bottom left people feeling dehumanised and forgotten during conflictual times.

Marginalisation: dimensions of deprivations

Iris Young writes that "marginalisation is perhaps the most dangerous form of oppression" and does not stop even when people may have adequate food or shelter. For ordinary people, marginalisation means being deprived of the ability to function in different areas because of not being acknowledged as a person with capabilities. Young describes the deprivation as "cultural, practical and institutionalized" (Young, 2011, pp. 53–55). The marginalisation I encountered against the Quechua culture was deeply rooted in education, from political to social participation.

Speakers from Quechua, Aymara and other original oral languages in Peru had a long history of being marginalised. Victimisation attributed a devalued perception to the cultural characteristics of women's identity. For example, the term illiterate was categorised as connoting second-class citizens (Henriquez Ayin, 2006, p.12). Through humiliation, stigmatised rural Quechua for being living in poverty are marginalised in their agencies across social, political, economic, and cultural sectors. Children were also widely marginalised and stigmatised for being poor, as the "poor child". According to UNICEF, 45% of the Peruvians live in poverty situations (Sinervo, 2013, p. 402), and "children in Peru living in extreme poverty are about 10 percentage points higher than the extreme poverty rate for the total population (16%)" (UNICEF, 2008). Stigmatisation changed in form and intensified during conflict, resulting in acute deprivations in relation to empowerment. Deprived of exercising agency, their unequal value perceived on the hierarchical identities resulted in some being vulnerable to being marginalised by others who felt entitled to humiliate. Disempowerment through systemic marginalisation became a structured form of oppression.

Education

People, illiterate in Spanish, voted for the first time in Peru at the same time that the war started (Henriquez Ayin, 2006, p. 13). Peru is a multiethnic country with 65 linguistic groups, the majority of which are rural, peasants, and/or Indigenous. Over 50% of the population have Quechua as their mother tongue, and women make up 73% of the illiterate (in Spanish) population (see more, points 9–10, 332, The Committee on the Elimination of Discrimination Against Women [CEDAW], 1995). It is concluded that the most marginalised group in Peru were peasants and/or Indigenous illiterate rural women. The system-built impediment for those not able to read and write, yet Oral cultures with encyclopaedic memories (Kelly, 2017), were set at the bottom of the educational hierarchy. One of Young's arguments about the "myth of merit" describes how performance evaluations are to justify discriminatory practices, putting at a disadvantage the marginalised minority. The "five faces of oppression" described by Young are plausible in terms of how a professional was defined based on the education received, which will require "acquisition of a specialised knowledge that entails working with symbols and concepts" (Young, 2011). The failure to recognise capabilities from oral languages is reflected in phrases in the work of scholars referring to the peasants kept in "credultity, passivity, and ignorance" (Piel, 1970, p. 118). Connoting specific characteristics of the peasants and/or Indigenous with negative remarks supports the association of ways of being to keep their identity with a sense of powerlessness and survival.

The consequence of mothers being deprived of education affects children's development and enhances their marginalisation. Already, children in the Ayacucho and its Andean region represent those living in poverty and extreme poverty and fall into the bottom of the category within the "Child Development Index" CDI (UNICEF, 2008). In the highlands and the jungle area, I noted the significant lack of access to schools in rural areas. Children must walk for hours to get to school. When the parents had limited resources to feed their children, the children often supported their parents by working in the land and harvesting. Like in many colonial projects, Indigenous people did not access a school system for centuries. When levels of indigeneity were hierarchically divided, those closer to colonial trends were the first to assist schools placed through religious missions where their Andean agency was forbidden. Oral agency and its outer manifestation through other artifacts were not recognised as knowledgeable, since being illiterate still has a negative connotation. Young talks about how the education system is structured to marginalise those who do not play a part in it (Young, 2011). Oral agency is not a disadvantage, and lacks acknowledgement from the education system and its structures. Learning to read and write the coloniser's language contributed to a higher place in the hierarchy in terms of education, in which a duality of education "versus" non-education are perceived with unequal values. Those with "education" are dominant and believe themselves to be entitled to oppress those with "non-education". Kelly (2016) points out, "for most of human history, there was no writing (2016, Location No. 606), and whenever Indigenous knowledge was usurped and replicated in literature, many qualities of knowledge were lost (Kelly, 2016; Yunkaporta, 2019)" (Rodriguez Carreon & Vozniak, 2021, p. 38). This does not mean that children in rural areas should not have access to reading and writing, yet to recognise that their agency and their ancestral oral technologies were not acknowledged, and over time, a collective embodiment of hierarchical ways of knowing created a perception of what education means.

Social culture

Despite Quechua, Aymara, and other ancestral cultures holding ancient wisdom, they have been oppressed since colonial times and stigmatised against their customs and belief systems, and, in general, perceived by the dominant system as a non-privileged culture. Deprived of value from what Freire (1972) called "all domination" involving "invasion". "Cultural invasion" served the invaders to penetrate their culture and ignore others' culture. "The invaders mold; those they invade are molded"; it is a static view of the oppressor, with a superior sense of their world (Freire, 1972, pp. 121–129). Young points out that certain groups were conditioned to oppression (Young, 2011, p. 40). Those who possessed the predominant culture within the country subjected people with oral knowledge to oppression through marginalisation. In the aftermath, people who felt alienated to Spanish trends tend to emphasise the need to be recognised as superior. Five decades ago Piel (1970) published, "peasants of all races created the conditions of incredible luxury and refinement for a creole society whose eyes continually turned towards Europe" (p.132). Historically structured, this segregation between urban culture with Spanish trends, and Quechua or Aymara cultures trends in the rural population, has remained prevalent.

Webb (2011) describes the economic values of the Andes within a particular community in Ayacucho. During the colonial period, the people from the Andes had a catastrophic psychosocial crisis during the sixteenth and seventeenth century. The Viceroyalty and Republic periods ignored the structural consequences of their oppression in subverting the people in the highlands and the jungle. Then, the Spaniards' descendants born in Peru were treated differently and, at the same time, distant from those native to the highlands or the jungle. Over time, people living in isolated areas were marginalised from several cultural ways of living without access to potable water and electricity. Even without these needs being met, people from Ayacucho enjoyed the culture that identified them (Webb, 2011, pp. 13–34). In particular the peasants from Pomatambo located in Ayacucho where Webb did his research. The people from the Andes maintained their own cultural rural life within the imposed limitations to access economic resources. Having the institutions geographically far away and being ignored by the oppressors became somehow a resource to protect collective cultural ways of being Andean. Othering as a culture, not caring and invisibilising peasants and Indigenous people's agency became colonial authorities' good fortune. For the colonial plan, the wider the gap with poverty, the "Other" were not intervening in their system.

Ways of being and doings in language, fashion, food, trade, and community organisation are some of the aspects of life relating to economic and political power. The alienation from Spanish ways of dealing with life was coercive and at the same time necessarily convenient (survival) if the people wanted to avoid marginalisation. As an adverse effect it perpetuated the disempowerment connotation of rural identity. Many felt inwardly wounded in their choices for changing their ways of being and doings on what they valued and had a reason to value to satisfy the dominant culture. During conflict, their cultural space to exert Quechua's capabilities was repressed. The Truth and Reconciliation Commission acknowledged that centuries-old cultural divisions did not disappear in conflict, but they did change forms. Being from Ayacucho during the war conflict was perceived to be a part of the conflict, and a part of those who created the conflict (CVR, 2004, pp. 174, 337), During the war, the dominant system stressed cultural marginalisation and ways of beings in this world as Indigenous and/or peasants.

Political

The conflict brought extreme insecurity for men in the Andes. As patriarchal as it was, man was the decision-maker in the highlands before the conflict. In conflict, communities were left without authorities and without committees to discuss and to decide their political questions. Rural women then assumed those positions and started to participate in meetings, committing to the village's common goals. This participation grew after conflict because there was not a political and urban public interest in what rural women were really doing. But, once their intervention grew in public spaces, those in power shifted their attention to them. Then, rural women were further repressed undermining their political agency and skills.

Women in the Andes had limited access, or none at all, to political venues in order to exercise their capacities before 1980. According to Isabel Coral Cordero, whom I interviewed in 2009, men in the Andes used to serve on committees to take decisions about village matters. In meetings men would normally sit in circles, and women would be silent at the back of the circle with no chance to access political power. At the end of the 1970s, when women were "hungry" for political participation, the Shining Path started to capture as sympathisers people with access to higher education, particularly from the urban areas of the provinces. The terrorist group noticed the absence of women's representation in politics and began to plan a gender agenda. They started to invite women to be part of their organisation. Women who were recruited thought there was a great opportunity to be heard because they were largely politically marginalised. They thought it was not patriarchal. However, as women came to be segregated to play traditional and domestic roles, they realised that the Shining Path did not really wanted women's participation in the group making decisions.

Institutional and cultural inequalities kept rural women at the margin of the political system. This does not mean they lacked agency in political matters. The marginalised participation was because the place for decision-making had not learnt oral culture values. Coral Cordero also said that changes to access political transformation remained unknown (personal communication, 2009). Peru was a stratified society with institutions that generated inequality which favoured distribution of power based on families with more economic access (Ludwig, 2011, pp. 98–99). The existing inequalities considered Andean women in their traditional roles as homemakers. And their participation in economic dynamics and in decision-making for the community remained the men's job until the conflict. Women were challenged to confront their communities' survival needs. With political agency, they transferred the capacities—skills learned in their homes and with other women in their community—to their public roles. The government did not acknowledge women's community responses in the aftermath as an agentive role during conflict.

Surviving: *passive agency*

Over centuries, the system's power relations legitimised oppression, leading to different forms of conscious and unconscious social agreements before, during, and after the conflict. This historically structured unequal system becomes more complex when identities interact in several spheres and within different contexts. During the 20 years of war (1980–2000) in Peru, identities and their ways of being and doing things were seen within a pyramid perception. Internalised oppression meant that some identities were Othered and rationalised as lesser humans.

Women's main priority was finding ways of surviving by seeking food and staying safe. Certain identities are prioritised in times of conflict (A. Sen, 2007, p. 27), and new identities emerge. With the emergence of their new identities, women connected with other women from different villages and, unified as a collective identity, exercised new roles. Still, to survive, identities challenged hierarchical roles and struggle with power relations. Affiliation with the same identity leads to a better outcome for the group, and the particular identity in association has the possibility to be extended (A. Sen, 2007, p. 23). Rural women lost their husbands as men were targeted by both the Shining Path terrorist group and the paramilitary extrajudicial forces, and their identities as wives and mothers became predominant, but not only in domestic spaces. As people from the highlands were displaced, forced away by the insecurity and impoverishment of their communities, their survival depended on seeking secure shelter, food for their children and information about their husbands. Again, when they were displaced, migrating their roles into a new form, they adapted their *survival agency*, weaving in *active agency* with resilience and capacities reinforced through their organised groups to empower themselves.

I found in the field that there were challenges to survive human rights abuses by the peasants and keep sanity at the core of their motherhood agency. In Ayacucho, for example, mothers started associating by one mother talking with another mother. Motherhood's agency awareness became viable through relating. Their capabilities to overcome common challenges were reinforced by acknowledging each other's courage. Similarly, when a widow was talking to another widow, their relationship as widows highlighted their primary identity. I contemplate the reason behind relating was to create compassion and empathy among identities, and relate as humans. Relating created strong links between women peasants to challenge the two threats to their human dignity, "survival and human rights abuses", as Coral Cordero mentioned (personal communication, 2009).

When confronting oppression, one needs "simultaneously reflecting on themselves and on the world, [to] increase the scope of their perception" (Freire, 1973, p. 55). During the conflict, women peasants' experience entered a moment of transition. The oppression was reinforced by aggressive and repressive measures, and reminded peasants of their connoted, disempowered agency within the system of domination. This period of transition, however, as Freire said, opens opportunities for *conscientização* (conscientisation) (Freire, 1973). Rural women became aware of their position in their system during the conflict and questioned their unconsciously performed roles through cultural lenses. There was more *conscientização* of their motherhood agency. But the continuing repression and threats against them when they demanded to know the whereabouts of their sons, produced uncertainty in their agency. Agency then fluctuated between *conscient agency* and *active agency*, not conscious but moving out from *passive agency*. When people immerse in their own world, the peasants do not notice that they are trapped with the "interests of the oppressor whose image they have internalized" (Freire, 1973, p. 38). It is survival. I considered here that the aim of the oppressor is to keep their victim's role in a state of uncertainty by discouraging their agency capabilities, in particular the identity with which the women themselves identified, their motherhood. For the oppressor, motherhood is not seen as an identity of strength, consequently in conflict, women's *active agency* is still invisible. The consequences of excluding the inner qualities of motherhood undermine the recognition of women's agency. Internalising the association of people living in underdevelopment with powerlessness supports the unconscious perception of hierarchical identities.

Stories from the isolated mountains: meeting underdevelopment

A mother's quote regarding one battle in Ayacucho at the end of the colonial period at the beginning of the nineteenth century mirrored the feeling of mothers in the aftermath of the twentieth century in the Peruvian conflict:

> My sons and the rest of the dead still belong to the owner of the horses and the owner of the lands, and the battles. A few apple trees grow among their bones and the tough gorse. That's how they fertilize this dark tilled land. That's how they serve the owner of war, hunger, and the horses.
>
> (Cisneros, 2005, p. 180)

Perceptions of social injustice which occurred during the twenty years of the Peruvian conflict have not changed the oppression since the battle of Ayacucho. By the end of the twentieth century poverty had not been improved through traditional development. For example, in November 2009, Rosalia (pseudonym have been assigned to maintain the anonymity), a hard-working woman from the mountains of rural Ayacucho, was one of the family members who attended a ceremony to receive the body of her husband killed during the conflict. The victims living in poverty killed in Santo Tomás de Pata in 1989 were part of the massacre perpetrated by Shining Path members who buried the bodies clandestinely. More than two decades later, the family members were waiting for justice—In Chapter 5, In the search for truth and justice, I shared the story of this community's grieving process in relation to transitional justice.

Rosalia was dressed as a typical rural peasant with her hat and *polleras* (pleaded skirt). Her physical appearance had a connotation of identity structured as being an Andean Quechua woman living in the highlands, an environment where the absence of the State is easy to notice. She was a grand-mother victim-survivor sitting in the area designated for family members of those killed in conflict, she had a tag with her name on it. Once I connected with Rosalia, I learnt she was a mother of six children, the widow of a person who held authority within her province, and she used to work in agriculture. The conceptualised "label" of victim-survivor is because she was tortured and saw her husband being killed. She survived the lack of access to basic resources and the violence in war by working as "anything", "working like a man"—in her own words—to provide food for her children to survive. As State services were not available for her, she was reminded of her condition in oppression during the conflict.

On the border between Ayacucho and Huancavelica lies Santo Tomás de Pata, one of the most impoverished communities in Peru lacked water and electricity, and proximinity to school and health centre. The precarious road between the Santo Tomás de Pata area and the capital of Ayacucho—Huamanga—are mirrors of the poor conditions of State development and human development in the highlands. The area concentrates the highest number of people living in poverty and with the highest illiteracy rate in Spanish among Quechua women.

Women like Rosalia show resilience built through exercising survival agency as a mother and active agency in her community. In a different context like the capital, Lima, her characteristics are missed to be perceived, internalised, and associated with social, political, and economic dimensions of agency. In the culture of domination's narrative, her identity is negatively connoted; she becomes the poor woman, the rural, the peasant, the Quechua, someone who has considerably been oppressed and marginalised for her condition.

Racism and disempowerment mechanisms have largely grown since colonial times against the rural people (people from the communities, peasants and/or Indigenous) and, through time, are now unconsciously structured systems of power domination. Poverty is manifested when those who feel different to Rosalia or superior to her, attribute disempowerment to her identity. The entitlement to perpetrate oppression due to the misconstrued perception of identities associated with negative connotations sees the system with a layer of fogginess that blinds the capability of sensing the human condition of the Other. Amartya Sen (2007) argues, in the eyes of aggressors there is not much we can do to persuade them "to take us to be different from (or more than) what they insist on taking us to be" (p. 31). Using Sen's viewpoint, in the eyes of the oppressor system, Rosalia is a disempowered woman without agency.

"People see themselves in many different ways; they have many identities" (Stewart, 2008, p. 7), depending on where they are located, their background, age, cultures, and so on. In Rosalia's case, even if her maternal and widow roles become known as part of her identity, oppressive views against her can co-exist. This negative perception lies in the core identities connoted as unequal in the Peruvian system, mainly in Rosalia's language affiliation with Quechua. When rural

Andean women interacted in other surroundings, like in the urban capital or places where they do not find themselves in harmony, their identities were perceived in association with a constructed collective experience of their marginalisation. In saying this, I wish to clarify that she may not sense that her identity has a disempowered connotation in a different environment—or among people who are not oppressors—in situations where, for example, to be peasant and/or Indigenous enjoys a positive recognition. At the same time, the oppressor in perpetuating marginalisation, exclusion or other forms to distance theirself from the Other compromises consciousness and, therefore, its connection with theirself *vital core*.

Exclusion: "An alien village within Peru"

All colonised countries share the same historical sourness of collectively embodied sense of exclusion. Centuries later, oppression and its forms of exclusion still stirred through Othering, affecting peasants' sense of identity. Before the conflict, people in the highlands used to say: Then, my village was there, ignored, an alien village within Peru, a phrase from Primitivo Quispe's testimony recorded by the TRC (CVR, 2004, p. 2). Primitivo was talking about how he felt about a system that did not recognise his community and culture, and he felt was not considered part of the country. The peasants felt aliens to citizenship. They felt excluded from the resources of the State. Exclusion continued when the absence of the State was replaced by the abuse of State during conflict. "For many women, problems of nationality, class, and race are inextricably linked to their specific oppression as women" (G. Sen & Grown, 1988, p. 19). Primitivo's statement shows us another face of oppression.

It is important to note that not all rural women survived conflict. They were all vulnerable to the risk of being killed by starvation, torture, bombs, and the displacement encompassed in the conflict. As Sen writes about those who have lived in poverty, it is sometimes just a matter of luck (A. Sen, 2007). When the violence started in Peru people with fewer resources, or no resources were the most immediately affected (CVR, 2004, p. 22). People without resources, mainly men, will escape once they knew they would be targeted. Women had sufficient awareness to know that their homes were not safe and found mechanisms to sleep outside their home, first, and then to opt for displacement. Displacement is not a decision with freedom. Women lacked human security capabilities from the system to be at home. They feared they would be assaulted.

Excluded from State development, basic services were not part of their consciousness. The lack of interaction with resources that was part of urban life conditioned rural women's capacities as being vulnerable. Excluded from interacting with "development" in their communities, the system in domination perceived peasants and/or Indigenous people as unknowledgeable, even though it was the institutions that were ignorant of peasants and/or Indigenous reality. In other words, the unfamiliarity with the mechanisms called "development" stereotyped people as uninstructed. Victims-survivors excluded from state development resources and from acknowledgement of their potential as agents in development needs a system with human-centred and consciousness approaches internalised, as seen in Chapter 2, *Consciousness-capabilities* under the umbrella of human development and human security.

Peruvians living in isolated communities knew their daily struggles during conflict affected their well-being. However, they did not have access to the system in domination to transform their needs into resources. The top-down development approach failed to acknowledge women's agency. Similarly, when conflict emerged in underdeveloped areas, top-down peacebuilding approaches failed to acknowledge cultural oral agency. The hierarchical categorisation of identities, which is not unique to Peruvians, has grown since colonialism, leading to a more significant gap in perception among individuals. There are more labels of identities, and also more racism and less humanness among them. The mountains and the jungle had a more antagonistic ontology than the urban areas, where the conflict was experienced differently. Urban people have conceptualised a different meaning to development and are more associated with economic consumption. During the conflict in isolated areas, access to proper roads to reach safety or to make demands with public servants in government institutions was a matter of life or death.

I want to take a *pause* here to acknowledge that there is also urban poverty, whose identity is stereotyped, depending on their access to economic resources and their ethnicity, and the armed conflict was also experienced differently than in rural areas and affluent urban suburbs.

During the war, Peru's social exclusion and poverty had a particular face: rural and peasant (CRV, 2004, p. 22). And, "where there are strict dichotomies that lead to destructive 'othering' and ethnic hatred, it is hardly surprising that harassment, violence and discrimination, allied with poverty and exclusion from full participation in social life, create fertile grounds for conflict" (Porter, 2007, p. 52). When conceptualising and materialising an Othering, the internalised collective perception depends on the narrative by the system in domination. At the same time, the system depends on subtle and direct communication coming from their pillars that make their system a structure that the collective relies on to belong. Internalising Othering risked further distance among Peruvians, facilitating the conditions to enable assumptions and undermining the peasant's agency in the power over narrative.

The absence of government

The absence of government institutions in the Andean communities signified the unreliability of its access. In traditional development, poverty is characterised by the lack of essential resources to meet fundamental needs, resulting in lack of nutrition, water, sanitation, housing and education (G. Sen & Grown, 1988, p. 38). For A. Sen (1999), development needs the abolition of representations of poverty, deprivation and the lack of opportunities by repressive states (p. 3). Thus, I highlight that analysis of those living in poverty oppressed through marginalisation are essential to generate inclusive programs for development (G. Sen & Grown, 1988, p. 23). When a government is perceived through the representation of its institutions, and its presence is absent in people's lives, marginalisation is defined in the eyes of those withholding power. For Andean women, excluding when their identities as peasants were segregated based on their appearance revokes their condition as oppressed, instead of connoting their appearance in gender, language, and ethnic identities as positive attributes. The absence of State institutions with basic services added challenges by increasing and intensifying vulnerability to rural *survival agency*. In Felicita's story presented in the previous chapter, her agency was disempowered by her son's illness. I found this lack of access contributed to marginalising her further in her freedom from fear. In the same way, in their search for security and survival resources, peasants experienced traumatic events that threatened with *llakis*. For example, in the case of Peru, the inner and outer ways violence reproduces itself is from exposure to what Johan Galtung called cultural violence(s). Galtung's (1990) cultural violence includes structural violence (lack of food and shelter), the horrors from direct violence (disappearance, torture, massacres, and others), and/or slow violence (racism, discrimination, marginalisation, patriarchy, and exclusion). The lack of services for providing support for meeting those needs added pain to the architecture of the invisible violence for Quechua people. Atherton (2003) explains marginalisation as a process that is not only complex but also has several layers and causalities occurring simultaneously. The unattended threats contributing to slow violence is the soil grounding of the unconscious realm in war.

Conclusion: otherness, the root to oppression for a blind system

The gap of unequal perceptions between oppressors and oppressed is a result of social, political, cultural, and economic dynamics reinforced by subtle and direct Otherness. Most often, what seemed to be the oppressed's passive response was pre-conditioned learning from former systemic repression. In this sense, oppression is embodied in both the oppressor and the oppressed and manifested as a power gap. Young used this language and analysed different kinds of oppression, one of them, marginalisation (Young, 2011). Atherton (2003) argues that although "poverty remains the commonest reason for people's exclusion from social flourishing, marginalisation is a much more multilayered and multifaceted phenomenon. It affects different groups, including sexual, gender-based, racial, and ethnic, even though relative deprivation is a recurring feature of them all" (pp. 57–58). In Peru, those who felt strongly identified with an identity negatively connoted, related to their gender or racial identities, were at risk of embodying and enacting the oppression that marginalised and disempowered them. It could be said that oppression is a slow hazard that is neglected, which occurs when people with a positive identity connotation interacted outside of their community structure and receive negative identity connotation. To find the cause of the problem, Atherton (2003) suggests listening to the people living in poverty. They are those 1.3 billion people on less than a dollar a day. To get the human sense of where justice will emerge to meet their needs (p. 59). Listening to *pause* (Rodriguez Carreon, 2023) allows one to see the unseen dynamics of slow violence within the system.

Living in poverty also means living deprived of a few things needed to make people feel dignified. Atherton states:

> Clearly, any consideration of relative deprivation has to acknowledge the presence and significance of absolute poverty as one of a spectrum of deprivation. As starvation, it is the ultimate marginalization because it threatens life itself.
>
> (Atherton, 2003, p. 60)

While Atherton states absolute poverty as the ultimate form of deprivation, for rural women being framed as poor was systemic perpetuation of deprivation. According to the CEDAW (1998), 44% of women are affected by poverty (p. 3). It obstructs human agency and its development. Being framed with identities that do not represent an economic and social "value" affected people's identity and internalised their marginal status in the Peruvian system. Wolf (1986) said, internalisation is a process, and where external realities become part of a person's subjective world, for example with unequal power, people from the minorities become participants of the oppression when felt oppressed. When segregation and marginalisation are interacting, structured oppression is forming and transforming, hiding in the subconscious. Their internalisation works passively on the outside, leading to structured power relations of identities of passive agents. Prolonged perpetuated throughout its various forms of oppression, when embodied, compromised their *vital core*.

Historical factors of lack of equal opportunities and empowerment that lead to oppression, dominium and marginalisation was found in the first Caribbean Human Development Report for Latin America (2010). At the same time, "traditional inequality, such as unequal treatment of women in sexist societies, or discrimination against members of other racial groups, survive by the unquestioning acceptance of received beliefs" (A. Sen, 2007, p. 9). In Peru, women have limited opportunities, positioning them at a level below the men (Guillerot, 2007, p. 41). The needs principle to foster a community for equality is tied to the creation of fostering collective solidarity and fellowship, in which certain forms of power, exploitation and humiliation end. "And in which vices of character where inequality is a constitutive condition—such as dependence, servility, sycophancy, arrogance and snobbery—are overcome" (O'Neill, 2005, pp. 74–75).

Structural inequality has prevailed in Peru for centuries, portraying the peasants as poor. In the novel called *Aves Sin Nido* (Birds Without a Nest) written in the nineteenth century by the Peruvian Clorinda Matto de Turner (1988) explains how oppression perpetrated against the Indigenous was represented. Inferior values were framed for those categorised as poor, and positive meanings were attached to the superior moral values of those in higher economic positions. Nevertheless, the stories show how some families did not marginalise those with connoted unequal identities and, with compassion, did not follow the choice to become an oppressor. At the same time, peasants who connected with beings and doings positively did not see themselves as oppressed victims. Yet, measures of disempowerment come from dominant alliances at the top where the peasant's culture was marginalised. For example, the mechanisms from the church and the government of disempowerment and marginalisation of Quechua are shown to be rooted in colonial times.

References

Adorno, R. (1986). *Guaman poma: Writing and resistance in colonial Peru.* University of Texas Press.

Atherton, J. (2003). *Marginalization. Part 2: On marginalization: The second essay.* SCM Press.

Burke, J. P., & Stets, E. J. (2009). *Identity theory.* Oxford University Press.

Committee on the Elimination of Discrimination Against Women CEDAW. (1995, September 20). *Consideration of reports submitted by states parties under Article 18 of the convention on the elimination of all forms of discrimination against women.* Third and fourth periodic reports of States parties, Peru. https://www.un.org/womenwatch/daw/cedaw/cedaw19/peru.htm

Committee on the Elimination of Discrimination Against Women CEDAW. (1998, June 22-July 10). *Concluding comments of the Committee on the Elimination of Discrimination against Women: Peru.* Supplement N 38 (A/53/38). Comments 309 and 310. Excerpted from: https://www.un.org/womenwatch/daw/cedaw/cedaw25years/content/english/CONCLUDING_COMMENTS/Peru/Peru-CO-3-4.pdf

Cisneros, A. (2005). The Battle of Ayacucho. In O. Starn, C. I. Degregory, & R. Kirk (Eds.), *The Peru reader: History, culture and politics.* Duke University Press.

Collier, P. (2008). *The bottom billion: Why the poorest countries are falling and what can be done about it.* Oxford University Press.

Comisión de Entrega de la Comisión de la Verdad y Reconciliación (2004). *Hatun willakuy* [The great story]: Versión abreviada del Informe Final de la Comisión de la Verdad y Reconciliación [Abbreviated Version of the Final Report of the Truth and Reconciliation Commission]. Navarrete.

Freire, P. (1972). *The pedagogy of the oppressed.* (M. Bergman Ramos, Trans.). Penguin Books.

Freire, P. (1973). *Education for critical consciousness.* Sheed and Ward.

Galtung, J. (1990). Cultural violence. *Journal of Peace Research, 27*(3), 291–305. https://doi.org/10.1177/0022343390027003005

Guillerot, J. (2007). *Para no Olvidarlas más: Mujeres y reparaciones en el Perú* [To do not forget them anymore: Women and reparations in Peru]. APRODEH, DEMUS, PCS.

Henriquez Ayin, N. Z. (2006). *Cuestiones de género y poder en el conflicto armado en el Perú* [Issues of gender and power in the armed conflict in Peru]. CONCYTEC.

Kelly, L. (2016). *The memory code: Unlocking the secrets of the lives of the ancients and the power of the mind.* Atlantic Books London.

Kelly, L. (2017). *The memory code* (1st ed.). Pegasus Books.

Lerner, G. (1986). *The creation of patriarchy.* Oxford University Press.

Ludwig, H. (2011). Desigualdad, diferencia y 'políticas de la identidad'. In J. Cotler & R. Cuenca (Eds), *Las desigualdades en el Peru: Balances criticos* [Inequalities in Peru: Critical balances]. Instituto de Estudios Peruanos.

Matto de Turner, C. (1988). *Aves sin nido* [Birds without a nest]. Peisa.

Mitchell, W. P. (2006). *Voices from the global margin: Confronting poverty and inventing new lives in the Andes.* University Texas Press.

O'Neill, J. (2005). Need, humiliation and independence. *Royal Institute of Philosophy Supplement, 57*, /December 2005, pp. 73–98.

Piel, J. (1970). The place of the peasantry in the national life of Peru in the nineteenth century. In *Past & present*, 46, pp. 108–133. Oxford University Press.

Porter, E. J. (2007). *Peace building: Women in international perspective*. Routledge.
Regional Human Development Report for Latin America and the Caribbean (2010). *Acting on the future: Breaking the intergenerational transmission of inequality*. United Nations Development Programme–UNDP. https://www.undp.org/sites/g/files/zskgke326/files/publications/rhdr_lac_en_2010.pdf.
Rodriguez Carreon, V. (2023). Liminality: Change starts within. *Challenges*, *14*(2), 25. http://dx.doi.org/10.3390/challe14020025
Rodriguez Carreon, V., & Vozniak, P. (2021). Embodied experiential learning: Cultivating inner peace in higher education. *Journal of Awareness-Based Systems Change*, *1*(2), 31–50. https://doi.org/10.47061/jabsc.v1i2.1179
Rotenstreich, N. (1989). *Alienation: The concept and its reception*. E. J. Brill.
Sen, A. (1999). *Development as freedom*. Oxford University Press.
Sen, A. (2007). *Identity and violence: The illusion of destiny*. Penguin Books.
Sen, G., & Grown, C. (1988). *Development, crises and alternative visions*. Earthscan.
Silverblatt, I. (1987). *Moon, sun, and witches: Gender ideologies and class in Inca and Colonial Peru*. Princenton University Press.
Sinervo, A. (2013). *No somos los pobrecitos': Negotiating stigma, identity, and need in constructions of childhood poverty in Cusco, Peru. Childhood, 2013*(20), 398. 21 May 2013. https://doi.org/10.1177/0907568213483598.
Stern, S. J. (2005). The tragedy of success. In O. Starn, C. I. Degregory, & R. Kirk. (Eds.). *The Peru reader: History, culture and politics* (pp. 124–128). Duke University Press.
Stewart, F. (Ed.) (2008). *Horizontal inequalities and conflict: Understanding group violence*. Palgrave Macmillan.
Theidon, K. (2004). *Entre prójimos: el conflicto armado interno y la política de reconciliación en el Perú* [Between Neighbors: the Internal Armed Conflict and the Politics of Reconciliation in Peru]. Instituto de Estudios Peruanos.
Theidon, K. (2013). *Intimate enemies : Violence and reconciliation in Peru*. University of Pennsylvania Press.
Thorp, R., & Paredes, M. (2010). *Ethnicity and The persistence of inequality: The case of Peru*. Palgrave Macmillan.
UNICEF United Nations Children's Fund. (2008). *Situation of children in Peru*. Executive summary.
Webb, A. K. (2011). *Nuestro propio sendero: Una comunidad andina y una economía de valores del mañana* [Our own path: An Andean community and an economy of tomorrow's values].. Instituto de Estudios Peruanos.
Wolf, C. (1986). Legitimation of oppression: Response and reflexivity. *Symbolic Interaction 9*: 217–234. https://doi.org/10.1525/si.1986.9.2.217. http://onlinelibrary.wiley.com/doi/10.1525/si.1986.9.2.217/abstract
Young, I. M. (2011). *Justice and the politics of difference*. Princeton University Press.
Yunkaporta, T. (2019). *Sand talk: How indigenous thinking can save the world*. Text Publishing.

4 War and its rawness through cruelty and repression

The armed conflict in Peru from 1980 to 2000 affected all the regions of the country, with major intensity reported in the mountains of Ayacucho. At the beginning of 1980, the Government returned to a democratic system after 12 years of being under a military regime. At the same time, with an act of violence by the terrorist group *Sendero Luminoso* (Shining Path [SP]) at a police station in a small village in the highlands of Ayacucho, the internal war started. The SP attacks escalated, and in the name of "security", the Government responded. A backlash against the Shining Path by the Peruvian defence administration escalated the warfare's violence. The Peruvian conflict ended with 69,280 deaths (Laplante, 2007, p. 314; Comisión de Entrega de la Comisión de la Verdad y Reconciliación [CVR], 2004). The Truth and Reconciliation Commission (TRC) (Comisión de la Verdad y Reconciliación CVR) reported that most people who were killed or disappeared were poor and illiterate and ignored from protection. They were seen as second-class citizens by the Government (CVR, 2004, p. 13).

According to the National Institute of Statistics and Information, one-third of poor women in Ayacucho were widows and heads of families (Theidon, 2004). Due to the fear of reprisal actions against them and their marginal status for those perceived in dominant positions of power, they were unable to place formal demands against those responsible for their widowhood, leaving those who committed the atrocities unpunished (Tamayo, 2003, p. 117). The irreversible experiences of Quechua women during the war are narrated through their stories of the 20-years of violence.

The war affected the most marginal[1] and excluded people throughout the country. Violent actions were also perpetrated within the urban areas (Youngers, 2003, p. 111), but with more intensity in the provinces. Around 75% of those who were killed and disappeared had Quechua and other native *linguas* as their mother tongues. Furthermore, around 68% of the victims did just primary school while the national average of people killed with primary schooling was 40%. Most victims did not receive secondary school education (CVR, 2004, p. 23). Men with children and responsibilities within their local communities were the principal victims. These Peruvians were battered by the terrorist atrocities perpetrated against their loved ones and, again, felt shattered when the military, police forces, and other state agents did not pay attention to their claims

DOI: 10.4324/9781003450955-6

(Youngers, 2003, p. 11). On the contrary, they were attacked by the military authorities (CVR, 2004).

Repression during armed conflict through violence affected people living in the margin of State development. Suppression was a tactic used by the Shining Path against those who did not follow them, and it was also used by the State forces to repress against anyone who matched their wanted "identikit". The resulting disempowered attitudes created an atmosphere of collective fear. People felt completely insecure. One of the testimonies exemplified empirically the lived experience under repression. A witness told how the military would invade their houses and take what was of value to them: their clothes, their animals, their food, and their people ... they spoke in Spanish, and people in the communities could not understand; they were frightening, and once the military members left, they went to look for the bodies of those taken away (Fernandez Castillo, 2003, p. 28).

How does direct violence threaten the empowerment of rural women's *vital core*? Compromising the capability to sense oneself affects the development of agency. Poverty is an intrinsic variable permanently present during and in the aftermath of the Peruvian warfare. It is carried as a sense of exclusion which existed in Peru before the conflict. According to Porter (2007), exclusion exacerbates the insecurity in people's lives. Living in poverty accentuated exclusion and stigmatised people for their identity associated with being impoverished. The "Peruvian conflict exemplifies the complex relationship of inequalities and political violence" (Suarez, 2011, p. 196). It does not mean that poverty is the main cause of the conflict. However, it is possible to conclude that the regions with a lack of resources were the most vulnerable and affected by the amount and intensity of violence (CVR, 2004, p. 22). The interdependence of both injustices, poverty and armed conflict, was a threat to their sense of security as humans when they attempted to exercise their freedoms.

Based mainly on the findings from the TRC's extensive report, I focus on the effects on the rural Quechua' *vital core* (sense of oneself) experienced by the violence inflicted every day. Repression, aggression, and deprivation during and after conflict periods left women's agency passive, without choice and tuned to survival. In the name of "saving" the population, both inflicted harms, the terrorist group Shining Path and the military. Their training and the implications of the inflicted violence led to their own dehumanisation. This chapter observes the main obstacle to the empowerment formation, namely, aggression and the survival stage of women's agency during disappearances and massacres. This period was accompanied by deprivations that rural women confronted, from mental health, body ownership, housing and community displacement. The extreme violence blocked the agency capabilities of the rural human development. Rural women found ways through informal efforts to actively build peace for themselves and those in their care.

Under the umbrella of security: human insecurity

The Shining Path and the military forces claimed to be in the rural areas for the people; however, they both failed and did not identify the real threat to rural human security. The Shining Path and the military members committed crimes and manipulated the discourse of freedom. A group of insurgents used terrorism with a sole indoctrination of obedience hidden under a discourse of fighting for *freedom from want*, or security from poverty. They were named by the communities "terrucos", which means terrorists. The reactionary military responded with tactics terrorising the whole rural community under the narrative of protection, to create *freedom from fear*. Freedom from fear represents human security and freedom from repression. Both used opposite dynamics to human security principles terrorising the rural population and isolated communities. How does the narrative of security filter grotesque violence, reasoning what happens in conflict when dehumanising entire rural and isolated communities?

Security here is defined as people's subjective experience, which is the sense of being free from fear. Subjectivity means "feeling" (Tadjbakhsh & Chenoy, 2009, p. 10). Human security refers to a focus on the *core* of human experiences as feelings. Human security is opposite to the emerged experiences from attacks made by the Shining Path and military to the rural people. There is a narrative to create the atmosphere that human security is based on human experience and its intentions (rational and emotional). But, as this chapter and this section illustrate about Peru, the "intentions" adopted by the Shining Path and the military did not match their "actions". In other words, it is not possible to achieve peace in human experience when this is sought by inflicting violence. For example, people living in poverty did not feel the lack of food would be resolved by building an army. Therefore, it is not possible to create human security based on freedom from fear when the core of the counterpart is inflicting fear.

In the Museo de la Memoria of Ayacucho (Museum of Memory of Ayacucho) is a painting with skeletons facing each other, one covered in red (the terrorists) and the other one in green (the military). It represents the dead found between the friction of both terrorists and the soldiers. The *terrucos* or the Shining Path members received many names, *terrucos* was mainly used in rural Peru and in the urban areas *terroristas,* both names resemble the English word, terrorists. They were also called the *senderistas*. The conflict of opposites can be understood through the theory of war. In his book On War, Clausewitz argued that "in war everything is uncertain … they [the military] direct the enquiry exclusively toward physical quantities whereas all military action is intertwined with psychological forces and effects. They consider only unilateral action, whereas war consists of a continuous interaction of opposites" (Handel, 1991, p. 6).[2] The opposites were the enemies, and in the Peruvian conflict were whoever "dared" to "contradict" either the *senderistas* or the military soldiers. The people in the highlands are then caught between "two fires", or two enemies. They would be killed either by the *terrucos* or by the military forces (CVR, 2004; Theidon, 2004; Henriquez Ayin, 2006; Youngers, 2003). This picture represents for rural women the feeling of being trapped between these two forces of violence (Figure 4.1).

The repressive dynamics of violence perpetrated were legitimised through policies particularly in Ayacucho. Repression was a threat to the population and provided an insecure environment resulting in people being displaced and abandoning their homes. To understand how a system in conflict established insecurity under

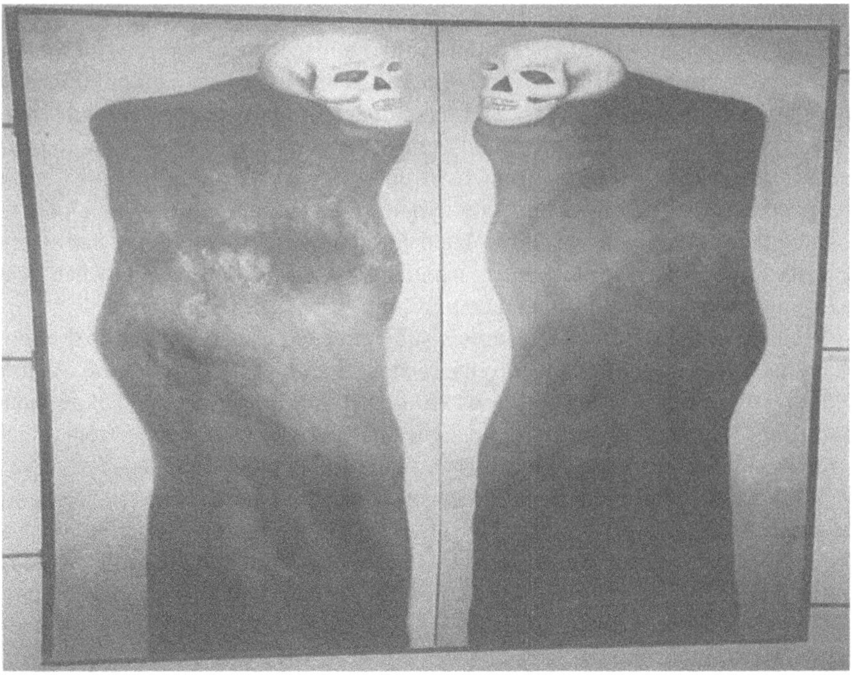

Figure 4.1 Trapped in violence.

Called *Las dos caras de la violencia política 1980–2000* (The two faces of the political violence 1980–2000). Photo (part of a bigger piece of art) taken at the Museo de la Memoria, ANFASEP museum, Ayacucho, Peru (Photo taken by Dr. Vivianna Rodriguez Carreon).

the umbrella of legitimacy, I analysed both groups' actions, the Shining Path ideology and its methods of *La Lucha Armada* (The Armed Struggle), and the army's response and the meaning behind their military training. In both cases the terror measures used against civilians undermined any possibility of protection for rural people. Their discourse and actions opposed human security yet used a narrative on security and its fear if not obtained.

The Shining Path: the ideological distortion

The Shining Path declared *La Lucha Armada* against the State in the small town of Chuschi (Ayacucho). This action was marked by the symbolic act of burning the ballot boxes just one day before the first democratic elections after 12 years under the military regime (CVR, 2004, p. 62). At that time, this event was taken by the authorities as an act of delinquency rather than a threat to the Government. However, later it would be found that this isolated action by the Shining Path was just the first of many constant attacks made against public institutions and the people.

Since 1980 and for the next two years, Shining Path actions focused on political propaganda (Theidon, 2004, p. 28). They created schools to win influence through their "pedagogic project" of Mao-Lenin-Marxist revolutionary ideology (CVR, 2004, p. 100). Persuasion and coercion of young men was a tactic used by the terrorists to recruit fighters. Shining Path's main discourse for their revolution was addressed as support for the poor and as against the rich. However, in contradiction to their discourse, people living in poverty were the majority of those who died during the war. It is noted that the Shining Path did not represent those they excluded, the rural people living in poverty (CVR, 2004).

The *senderistas* practised their ideologies by delivering their discourse walking through the difficult geographic and remote rural communities. They recruited those who followed their beliefs, abducted those who did not want to follow them, and killed those who opposed them or drastically rejected them. According to the CVR, these methods were to build fear among the *campesinos* (peasants) as part of their terrorist strategies. As explained in Chapter 1 The point of departure: Armed conflict in rural areas of Peru, the victims were mainly young people within the workforce (CVR, 2004, pp. 42–52).

The Shining Path's "bloody time" was initiated through various terrorist attacks which included first advocating *ajusticiamientos*, which means people taking justice into their own hands or public justice. However, the *terrucos* public executions' purpose was to build an atmosphere of fear, and they forced unwanted *ajusticiamientos*. For example, if a crime was committed within a community which could not be brought to justice through the government's traditional system due to the absence of their institutions in rural areas, then the Shining Path took as an advantage to show they were able to apply justice. The consequences of the ineffective judiciary system for the rural people were exploited by Shining Path members through indoctrination and action. The terrorists used the tactic of selecting those who were suspected of the crime, for example, for being known to be dishonest, then imposing extreme violence as punishment. Such actions were subsequent to the *Inicio de la Lucha Armada* ILA (Initiation of The Armed Struggle), the "bloody time" that was marked by massacres against the rural people and bombs to any institution that represented the power of the State.

According to the TRC report the extreme violence perpetrated by the Shining Path for minor crimes terrorised the people in Ayacucho. Overtime, people living in the highlands started to question the purposes and actions of the Shining Path. The public executions were taken by the *terrucos* as an easy way to eliminate anyone suspected of being *soplones*—those helping the military, named as traitors by the Shining Path members. The terrorists applied extreme violence for infidelity, minor robbery, or simply those who were pointed out as enemies. These inhuman acts perpetrated as punishment were rejected loudly, in particular by the women. For instance, "the women asked that if they were going to kill the fathers, they should kill the whole family". The *campesinos* strongly rejected the Shining Path violent actions that contributed to the SP own's failure and its overall decline (CVR, 2004, pp. 348–349).

As people in the highlands felt unsecure and unprotected, the rural civilians organised themselves in *rondas campesinas* (peasant rounds) with sticks and knives. Gradually they did not fear anymore and "the intensification of the *senderista* repression reflected internal problems and the weakening of its forces" which the Shining Path never acknowledged. Since 1984, even sympathetic peasants of their freedom from poverty discourse declined to follow the *senderistas*. By 1988, the peasantry were properly organised to defend themselves and called *Comités de Autodefensa* CAD (Committee of Auto Defense) (Del Pino, 1998). Legally incorporated into the Citizen Security System in 2022 is a Self-defense and Rural Development Committee. The CAD originated with the *rondas campesinas*. These were in part supported later by the Government armed forces (Coral Cordero, 1998, p. 358). They played a critical role in defending their communities (See more, Starn, Degregory, and Kirk, 2005). Yet, they were seen as the only choice to protect their communities during the war.

Failing to protect the peasants and Indigenous people from the Shining Path, the extreme violence reached two peak stages in the rural areas, in 1983 and 1989. The highest point of attacks in the capital, Lima, was in 1992 when the violence spread throughout the country (CVR, 2004). From 1990 the *senderistas* attacks were concentrated starting in the *conos* of Lima (CVR, 2004, p. 174). *Conos* are known as suburbs which due to internal migration were created at the outskirts of the city. After the *conos*, the violence reached the main economically "privileged suburbs". In Lima, car bombs started to explode in different districts, shopping centres, restaurants, and hotels.

The constant attacks in the capital subsequently increased the pressure on the government. Between 1990 and 1992 a curfew was established in the capital, and the constant blackouts were a frightening reality for the urban civilians. The insecurity prompted the population in urban areas to speak out loudly and peace protesters went to the streets (Henriquez Ayin, 2006). During the conflict, according to a leading newspaper report in 1992, 64% of the people agreed with the death penalty for the terrorists (Henriquez Ayin, 2006, p. 48). The leader of the Shining Path, Abimael Guzman, and his top followers were captured on 12 September 1992 and from there the Shining Path slowly reduced its activities (CVR, 2004).

The violence affected the country socially, culturally, politically, and economically. Community structures were broken for the *campesinos*. After the conflict, rural women were found to be those resilient and traumatised victims-survivors who were the principal witnesses to the violence. With a broken community, their families were disintegrated, their cultural customs were dispossessed, and their economic trading activities were vulnerable or disappeared.

Pause

I want to make a *pause* here to bring up a historical component from colonisation which was explained in the previous chapter, the concept of hierarchy among identities. Education was a survival to elevate the "race", and to be educated meant to be and to do things as Spaniards. What the groups fighting, the Shining Path and the system in power of domination, failed to acknowledge is the human flourishing knowledge from oral Agency, peasants, and Indigenous people. What knowledge means, it was also colonised.

The *obligated mind*: the Shining Path indoctrination

Ideologies indoctrinated through disempowered relations are opposite to the concept of education. The SP indoctrination in the name of "education" began in 1980 to recruit the youth to deliver "knowledge". The absence of education in isolated communities was exploited through discourse with repressive tactics to increase their propaganda. It was not the form of education as open learning based on an interactive process that enhances conscious knowing. The first SP members were mainly teachers, university students and professors with very little influence from the peasants' communities living in extreme poverty. However, over the following years, the indoctrination intensified, and children with just primary and secondary schools living in poverty were forced to be recruited to be key SP members.

At the beginning "power seduced [these] secondary students" into a "myth to progress" (Degregori, 1998, pp. 128–139). The Shining Path promised to end the pain of "poverty, inequality and marginalization" (Del Pino, 1998, p. 176). A new state was promised where the people would be represented and be part of the decision-making process. Arms were given to students so they could see physically how this power worked. At the start youths were given paints for graffiti propaganda and dynamite to experiment. They admitted doing it just "to blow it up" (Degregori, 1998, pp. 128–139), to see what happened. It was like a war game, child soldiers describing how the arms were given like "candy" to test. In the documentary "State of Fear, the truth about terrorism" by Pamela Yates (2006), one of the survivors recruited and forced to be a child soldier said that it was like a when a child is offered a *caramelo* (candy). It became vicious. The "Shining Path made a show of displaying its coercive capabilities to the peasantry" (Degregori, 1998, p. 131). If the child and youth soldiers were not "seduced" then they were obliged to dehumanise themselves by inflicting violence on others.

The youth felt confused. A psychological symptom that is recognised as trauma. They did not recognise themselves in fragmentation. With the being + Self connection broken = their *vital core* was compromised. Dehumanisation is an unconscious out-of-body experience. A young Rwandan woman narrates her loss of identity, soul, agency, and recognition from her lived experience of dehumanisation, when the body becomes "objectified, fragmented … and emptied of 'human' subjectivity" (Oliver, 2011, p. 90). When the body is gone through a process of objectification, it means the Self and the body are disconnected within the body field.

The people in the Andes were let down by the oppression first, and then their "illusion" of freedom was suppressed with inhuman dynamics under repression. Young peasants who were called on by the *senderistas* recalled the time at the beginning of conflict as an "illusion", for having thoughts on how they might free themselves from poverty. But the Shining Path members did not aim to free, but rather to manipulate the youth to think like them, by exercising coercive teaching based on repressive methods where the students needed to remain unconscious or ignorant participants without questioning or arguing. In a moment of transition[3], they became aware enough to decide to oppose the SP ideology, yet at the price of punishment. A feeling of superiority dominated the "teachers" doctrine over the young peasants. The SP members did not acknowledge what I would call the "natural knowledge" of the rural people's surroundings.[4] The *senderistas* felt superior and treated the *campesinos* as ignorants. The young peasants felt their minds obligated to survive the simultaneous oppression and repression of their wills. The Shining Path then started losing their members. They soon realised that they needed to keep the population illiterate, using not just fear but violence to induct the Quechua people (Del Pino, 1998, p. 172).

The rural peasant women's natural knowledge of their surroundings was soon disempowered by their sons whose indoctrination portrayed their mothers as ignorants. However, rural women's instincts made them want to protect their sons from becoming part of the indoctrination used by the Shining Path. The *senderistas*, at the same time, actively intensified the violence in their recruitment of those who were opposed to their ideology. As the following quote shows how the *Ayacuchanas* women were discouraged by the conflict:

> The first signs of peasant discontent emerged when children were drawn into the war. Mothers recalled that their children "were forced against their will, whether they wanted or not, they showed them arms, knives, spears; if you don't accept it you'll die". The women might collaborate or accompany their children in actions, but "the fear is only for our children, for what they can take away from us". It was from the mothers, who appeared most sensitive to the suffering of their children that the idea of resistance began to develop. The children's fate would justify the claim that Shining Path "nos ha engañado" (has deceived us). These words would be spoken insistently.
>
> (Del Pino, 1998, p. 171)

From this quote, I analyse the mothers' survival experience and how they gradually woke from Shining Path repression. Resistance was in the case above the core value protector, the primary motivation. This does not mean they did not have empowerment capabilities. The capabilities to operate consciously were impeded indeed. However, internally their resistance was part of their empowerment formation, and their agency within a passive and survival stage—passive on the outside, however, inner qualities move inwardly. They are learning ways to resist, and as long as their *vital core* was not affected, empowerment was in evolution. Noted by Ponciano, the mothers considered not having any more children because they did not want to see their children going on to war or going against their moral values (Del Pino, 1998). Rural women felt the war threatened what they valued and had a reason to valued. In other cases, repression undermined their core. Freud's (1989) analysis from the psychology of consciousness to the meta-psychological description of the process of repression states in Gay Peter's (1995) reader, "the essence of the process of repression lies … in preventing it from becoming conscious … everything that is repressed must remain unconscious" (Freud & Gay, 1995, p. 573). The *vital core* of many victims during the Peruvian conflict became unconsciously and gradually affected by the cruel unhuman actions. The terrorists made children watch how "they slit his throat with rope, the chocked him, pulled him apart, in front of everyone …". Sometimes they forced a close relative to commit the execution. These actions were during the "people's trial" when the peasants were deserted and the *senderistas* wanted to show the consequences to those who escaped (Del Pino, 1998, p. 185).

From 1984 the Shining Path conscripted entire families from places where they prohibited the practice of social and family customs, stopping the children from accessing their parents' care, and leaving them to survive by their own means. Later on, the SP would call these children "camaradas". On the SP campus, children were obliged to repress their feelings when exposed to direct war violence, and from an early age were indoctrinated to "kill or die". Consequently, in the attacks, they will go first, "screaming, burning, and looting". Del Pino writes how one felt obligated: "they gave me the knife … I felt I was in a dream, a nightmare, I became inert while trembling inside, … I went to him and said forgive me, and I gave him a stab in the chest and I screamed out of fear … Then in the stomach, in the heart, while I ended up smeared with blood and not understanding what I have done … I wanted to go crazy and escape that very moment" (Del Pino, 1998, pp. 185–186). People recalled that they did whatever the *terrucos* wanted. It was carnage, I had written in my notes, at the time of my fieldwork.

The cruelty escalated in combat as part of the militarisation of the terrorist group. Their divisions were hierarchical and racist. The discourse of the SP over the ideologies was over epitomised and repetitive to enrage *la gente de las masas* (the people from the masses) as *senderistas* used to call the rural people. As noted, they used to exclaim that the future now lay in the use of tanks and rifles! The armed revolution has started! Glory to the Marxism-Leninism-Mao Tsetung! (CVR, 2004, p. 103; Gorriti, 1990 p. 67). The narrative was aggressive.

Was it possible that after forced indoctrination through this repression the *vital core* of the population stayed *sano* (healthy)? It is noted that the Shining Path practised extreme *machismo* against those recruited; if they showed any feeling, like crying, they would be killed (Del Pino, 1998). This type of repression can affect the very core of the human, preventing empowerment formation. The insecurity in their vital needs was severely worsened by the Shining Path. Many did not survive and, as Del Pino noted, some died from disease and malnutrition (1998). Such repression was exercised after pressure, as Freud would call it "a proper repression" (Freud & Gay, 1995, p. 570). However, some deserted and prevented repression by indoctrination before their core was consumed completely: "in my inner thoughts I was always refusing, [I was] totally exhausted, oppressed; for this reason I've also left", Gliserio said (Del Pino, 1998, p. 183).

Conspiring for violence: the military

Transgressions against the collective Other in conflict inflicting physical and psychological violence is a repression measure at a political level. These actions inwardly obstruct the empowerment formation of human agency. How did the military use repression in the name of security? What experiences transmitted a sense of security to those whom they meant to protect? How were soldiers trained to exercise the duty of the military, particularly those who were involved in mass disappearances and killings in the Peruvian conflict? I quote here Nolen Gertz's repression analysis of Freud's argument about the consequences of war from the perspective of military training:

> It would seem that in the conflict between the urge to act on one's aggressive instincts and the demand to respect a ranking officer that a battlefield would offer the soldier many opportunities for releasing these "suppressed impulses" on the enemy. In fact, it would seem that it could be said that much of the purpose of military training is indeed to channel these urges in this very way.
>
> (Gertz, 2010, p. 68)

The instruction for discipline here shaped suppressed individuals. A human being who is going to inflict violence in his/her self will also develop violence in the Other—being the Other: the Self, the Social, and/or the Environment. Suppression is a disempowered inner action motivated by violence. Higher military ranks included violent actions in their operations orders to the soldiers at lower levels. In the case of those of lower ranks, they built an internal justification of their being in order to not become conscious of the inflicted action. Oswald Garrison (1889) argues that "universal military servitude is far too great a price to pay for insuring peace by any free peoples. Its dangers, its contaminating effects, the terrible weapon it forges for rulers, its reducing to men to a dead level, … [are] not meant or intended to increase efficiency in civil life. Its primary purpose is to turn out killers, not workers. It often destroys those it would benefit" (Garrison, 1889, p. 7). The expression of suppression is a violent weapon to the experience.

Latin America has a history of using military force against the Other by inflicting severe torture, kidnapping, and killing. Over the last decades such tactics were learnt by the Latin graduates in militarism at the "US Army School of the Americas".[5] For example, "the three highest ranking Peruvians' officers … were all graduates" and were convicted in 1994.[6] Former military students applied the teachings to regenerate the military services in their own countries. Many of those graduates committed the most atrocious human rights abuses. For example, Salvadorian death squad leader Roberto D'Aubuisson, Guatemalan General Hector Gramajo accused to be the architect for the genocide of Indigenous people, former Bolivian dictator, among others. They were "some of the worst perpetrators of human rights" (Gill, 2004). As Quigley said, they were accused of "massacring … drug trafficking, summary executions, leading death squads, and torturing" (Quigley, 2005). How can violence attempt to be justified with concepts? Freud's analysis directs us into thinking that by inflicting physical and psychological violence as part of the training, the recruits would then mirror what has been done to them. Freud reflects Einstein's words when he writes, "that it is so easy to rouse people's enthusiasm for war, and assume that something is at work in them that is receptive to this indoctrination—an instinct for hatred and destruction" (Paret, 2005, p. 16). A part of Freud's response to Einstein on the topic, 'Why war'? I analysed the course learnt by the military in this school to understand the unconscious dynamics—absence of the mind, indoctrination sin—of the systematic violence affecting the *vital core* by repressing their freedom capabilities.

Adding to the previous question, how does the public digest a discourse polarising the Self and the Other to justify violence? The answer may lie in the social perception of discourse represented as a goal rather than adaptive behaviour questioning the process of reaching the goal. For example, an article in a Peruvian political magazine detailed the training of the military. To enrol in a course for the armed forces, soldiers had to pass a rigorous exam. The meaning of rigorous in the article was portrayed as courageous. The action to confront was a very disciplined training to survive in combat (Caretas, 1983). Both training and survival have a social positive connotation associated with strength.

The Caretas' article, which analyses the military training of those going into combat in the rural regions, does not describe how the soldiers were going to "protect" the vulnerable people in need—the rural Quechua women living in poverty. Instead, it describes how soldiers were dehumanised in order to dehumanise others. The training was a physical resistance journey of discipline which included following orders, such as no more than three hours of sleep otherwise punishment was applied. The training of two weeks involved running between 25 and 40 kms, and jumps called *salto del tigre* or "tiger jump", one of the most dangerous actions. Another was a jump of over four metres and then falling into *la telaraña* (the spider's web), a net of ropes after an explosion. Consequently, a lot of the potential soldiers had accidents; half had fractures and had to abandon the course. Indeed, five people died during the 23 years of training at the school. Operations were also conducted in dangerous areas involving survival experiences, communications on how to infiltrate, sabotage, use of explosives, and other activities (Caretas, 1983).

This training meant for the soldier to accumulate the repression of aggressive instincts. Nolen Gertz addressed the relevance of how the soldier's mind is disempowered for no other reason than repressing his feelings of anger, and after releasing those feelings (as Gertz notes on Ernest Jones 1921), in combat. The military gives the soldiers the licence to "indulge" and engage when they are back for the "violence of combat". He notes:

> Though he may have been socialized to consider stabbing or shooting a stranger to be savage, vicious act, the soldier discovers that when on the field of battle – even if only on the practice version – such brutality can be seen as "permissible and laudatory", and if carried out in the right circumstances has been described as "heroic".
>
> (Gertz, 2010, p. 72)

The narrative of war is perhaps responsible for how militarism is portrayed as heroic even when committing barbaric acts. The article in a Peruvian magazine emphasised that we have seen the militaries, they look tired to pass tests that not just takes you to the limit of physical resistance, but it requires high skills, professional capacity and above all courage (Caretas, 1983). The training to become a soldier and to obey your superiors requires repression of your instincts by obstructing your feelings of anger (Gertz, 2010), and other disempowered dynamics. So how then are the soldiers' actions "heroic" in the narrative? The narrative intends to conceptualise disempowered and, at times, dehumanising training inflicted against the *vital core* of the trainee with a positive connotation for power. Inwardly, the being represses their feelings and accepts the intentional violence as valid for other means. For example, one can argue that the rigorous program created by The School of the Americas intentionally shaped its graduates so that they would commit serious atrocities like massacres, torture, and systematic killing, creating schools of how to preserve this system in order to disempower the "other" (see more, Sen, 2007, p. 172).[7]

It is explained here that the training used to disempower and dehumanise the "other" was made in essence—it is worth repeating—to disempower and dehumanise their recruited being, and their capacity to sense oneself. The training was conducted under the umbrella of being legitimate by "enabling" and "allowing" the act of violence. The violent actions were done in the name of security, yet by inflicting human insecurity. The suppressed soldiers in training lacked the capabilities of freedom to pursue what they value. And instead, they will follow up training with the purpose, which Gertz noted in Jones, of "extreme punctiliousness" of military discipline to commit "cruel acts". The order of the "Commander" and "the military is intended to allow the soldier to believe that he commits violent acts for the sake of the nation" rather than for a desire that was stimulated and is released at the time of conflict (Gertz, 2010, pp. 72–73). It is a "Gilligan's humiliation" style, like when someone who is dissatisfied with something becomes afflicted by the need to exert violence, and the person perpetrates force against someone defenceless (Keen, 2012, p. 196). Militarisation gave impunity when perpetrating violence under the umbrella of legitimacy.

The narrative and discourse show the job of the soldier as valuable. However, this is an area that has been not much researched (Gertz, 2010). Also, it deserves a systemic evaluation of how discipline works in the social context. Nevertheless, the value of the analysis here is the interpretation of the consequences of training and the effects that resulted when these were applied. The importance is the training dynamics for its influence on the disempowered formation against those who were not recognised as fellow humans during the Peruvian conflict. It is confirmed that people—their fellow humans—from the highlands matched an "identikit" profile as poor and rural (CVR, 2004). And, it is noted, that the trainers who went to prepare the soldiers deployed in Ayacucho's bases were from the "elite" group that practised their tough teaching at the School of the Americas (SOA). The head of the Peruvian academy for the military was formed at the "controversial" SOA[8] and has been involved in several massacres in Ayacucho where children were among those executed. The preparation was based on inflicting pain and violence on soldiers, then releasing them to perpetrate violence in combat. "If you look at the course ranges that are offered to these individuals, they in fact are a dedicated way of teaching military leaders in foreign nations how to subvert their local communities" (Kennedy, 1998). The inquiry here is how the collective consciousness will engage more with the courage side for enduring pain and strength, and other collectives will relate more to training that demoralises what it means to be human.

According to Freud's "process of repression", "the ego notices that the satisfaction of an emerging instinctual demand would conjure up one of the well-remembered situations of danger. This instinctual cathexis must therefore be somehow suppressed, stopped, made powerless. We know that the ego succeeds in this task if it is strong and has drawn the instinctual impulse into its organisation. But what happens in the case of repression is that the instinctual impulse still belongs to the id and that the ego feels weak" (Freud, 1989, p. 779; Gertz, 2010, p. 77). This action limits freedom by inflicting bodily pain/injury and/or killing (Davenport, 2007, p. 35). During the preparation for fighting, those who were combatants suppressed feelings and embedded the exercise of force to later release such repression onto the powerless. On the other hand, to survive, the powerless suppressed feelings against the perpetrators at the time of violence and, at the same time, tried to pursue the release of those feelings in their fight for justice. The following section portrays an example of a scenario where most of the repression was committed during the Peruvian conflict, under the State of Emergency.

State of Emergency

On 12 October 1981 the Peruvian Government declared a State of Emergency in seven provinces of Ayacucho and sent their special police forces called the *sinchis* (CVR, 2004, p. 115), a US-trained "counterterrorist" force. These defence forces detained in some operations the SP members; but it was also discovered that their intervention instigated several human rights violations. Women were raped and tortured by the *sinchis* (CVR, 2004, p. 246). Several stories given to the TRC evidenced the horror of the situation for women's vulnerability, and the violation to their human dignity during State of Emergency.

Between the years of 1982 and 1986, an escape by more than 200 prisoners organised by the Shining Path prompted the Government to conduct the "militarization of the conflict" (CVR, 2004, p. 247). This meant the military was entitled to, and possessed legitimate power to take control of Ayacucho, and other mountainous states under a State of Emergency. By this means, arbitrary detentions without a trace were perpetrated against anyone who was suspected of terrorism. Civilians of rural Peru, especially men, were exposed to the consequences of the State of Emergency such as rape, torture, disappearance, and extra-judicial killing.

This repression can be legitimate or illegitimate, and it means the exercise of force by the State (Davenport, 2007, p. 35). Davenport's definition quoted below exemplifies the exercise of repression used by those given authority to exercise power over those who were in a vulnerable situation.

> [It is] repressive because of who uses it and the influence that relevant behaviour has on the human body as well as mind. Indeed, perhaps the best way to conceive of repression is simply to think of it as a "power attempt"

> (an effort exerted by political authorities toward a specific end that relies upon physical force) … repression concentrates on the coercive behaviour of the state, research on social control highlights non state actors and so is deemed quite different. The focus means process of repression is important because it unifies the literature. Without this orientation, it would be difficult to imagine bringing together research on arrests, imprisonment, surveillance, disappearance, political bans, pepper-spraying, the closing of a newspaper, and mass killing.
>
> (Davenport, 2007, pp. 36–37)

This wide exercise of repression was not possible without some umbrella of legitimacy. During the conflict, the declaration of a State of Emergency meant in a way the beginning of repression, or the legitimacy of the exercise of repression. Some of the abuses perpetrated by the military against the less protected included abduction, house invasion without a warrant, and disappearances without a trace. State abuse affected the *campesinos* (peasants) living in marginalised exclusion.

The State of Emergency meant a curfew was imposed. During the time of curfew there was no law to protect the people. Houses were violated and extreme violence was perpetrated by the military forces. The rural population was not the primary and ultimate end which they aimed to protect. However, they were used in the discourse of State protection. This clearly intensified the "bloody time" in Ayacucho, and the rural dead were appraised as "casualties of war" (Theidon, 2004). The State of Emergency increased human rights violations with deprivations where rural civilians felt obligated to hide and leave their communities. They lost the freedom to meet, the freedom to circulate, and the freedom of private property.

The definition of repression includes, as Davenport argues, that individuals are killed, beaten, humiliated, raped, and tortured by a coalition between state, individuals, and militias against the State's proper purposes. It is necessary to look further at repression to revise the history of the place, the socio-economic context and the general structure of the government (2007, pp. 34–35). Revising the background of the victims we can summarise that the rural population who were oppressed were now under the authority of those who had ignored them for centuries.

Aggression against the *identikit*

The true enemies of the military and the State were the members of the Shining Path. However, the military did not have an acquired *identitkit* image, they could not know who the *terrucos* were exactly. The enemy, as Degregori argues, was stigmatised for their race, an image produced … an identikit of the violent person was equal to a being from Ayacucho, from the mountains, young, educated, Quechua = terrorist (Degregori, 2003, pp. 17–18). The danger of this lies in relying on the identification of physical appearance, on what people looked like; because anyone who matched the "characteristics" of being a possible *terruco* was seen as suspicious, and therefore guilty, and could be abused by the armed forces.

A former head of the military, Luis Cisneros, said in an interview which was transmitted publicly: "the police don't know who the *senderistas* were. We are going to kill 60 people in one village and probably only there were going to be three members of the Shining Path and the others were going to be civilians of this war". It is possible to conclude from this quote that the peasants from the village were not considered to enjoy full citizenship when their state had the responsibility to protect them from Shining Path members. Civilian young men living in rural or urban provinces matching the *identikit* were in danger. It is confirmed that the military was responsible for 60% of the people who disappeared (Reyna Hidalgo, 2013). The young population within the community were recruited by the SP, and broke the community's traditional structures; or they were disappeared, and subjected to extra-judicial killing by government forces.

Aggressiveness against those who matched the *identikit* image was an expression of misanthropy. It is a hostile dynamic that happens in every war. Those who were not killed, by disappearance or massacre, survived in many cases rape and torture. The experience of aggression during repression was embedded embodied as injustice—perpetrated under the umbrella of impunity. The rural population in this case felt constantly abused by systematic oppression, by *terrucos* aggression and by State violence. The experiences affected the core of their being, tearing down their feeling of being human. The *identikit* produced visceral experiences of fear with particular inner qualities of affliction for victims-survivors.

Forced disappearances

Disappearances are used as a social and political weapon for terror. The terror stays in the social psyche as a collective trauma. According to a recent International Committee of the Red Cross ICRC, article (2019), in the last decades, "it is estimated that at least 83,000 people have disappeared in Colombia, more than 20,000 in Peru, over 45,000 in Guatemala and more than 5,000 in El Salvador in connection with the armed conflicts. Globally, the historical trace of this tool for repression has been documented, as has its use in Latin America (Aguilar & Kovras, 2019). The threat of disappearances, however, is beyond the numbers given. What is the inner quality of experience threatening the capability to sense oneself in survivors when they search for those who vanished in the political unaccountability of the conflict's disappearances? The experience of having someone who disappeared left their survivors an emptiness. The psychological implications for the survivors can be overwhelming. "More than anything, they want to know what has happened to their loved one. It becomes an obsession and a constant source of great anguish … their pain does not fade over time; instead it lingers on, sometimes in silence, ready to resurface at any moment. With such uncertainty, they are not able to begin the mourning process" (ICRC, 2010). In the case of the late

Mendoza de Ascarza, tranquillity ended the day they abducted her son. It was too much pain … so she occupied herself with the search for him. She remarks:

> I don't forget that thought, always I am concerned in my head: How is my case going? Where is my son? Where are all the disappeared people? What happened to them? When are we going to know the truth? That is my concern.
>
> (A. Mendoza de Ascarza, personal communication, 2009)

The ANFASEP founder was in despair for the cruel reality. She represents the rural women of Ayacucho and their experience of uncertainty. The phenomenon of being in the unknown from the disappearance of a family member aroused inner qualities of uncertainty. It is a subjective experience in women's minds; a feeling in which, as mentioned before, lies the focus of human security. An obstacle that puts at risk directly the *vital core*, it is a threat to the consciousness of their agency and holds back the empowerment formation. Disappearance for the rural women can be described as a horrific atmosphere which prevents them from having any peace of mind. As Einstein noted, during conflict there is the symptom of "severe psychic weakness in humanity" (Paret, 2005, p. 15).

In wartime, emotions became numb and internally something shifted, for both perpetrator and victim. In other words, the inner Self changed. Interactions among people changed and transformed qualitative dynamics within relationships. Violent actions were perpetrated without awareness of experience. For most of the conflict, there was no noticing of the whole, no consciousness, so there was no perception of what was happening inwardly as a society. The narrative was indeed affected by propaganda, indoctrination and training received by *senderistas* and military members, as well as by invisible violence.

In Peru, male were tortured, disappeared and/or killed. Their houses were invaded. Their rights were violated. Women's bodies were violated. Their dignity was threatened. Disappearances of women's husbands and sons changed women's roles within their community. According to the Truth and Reconciliation Commission, the military forces used three mechanisms against the *terrucos* through a systematic disappearance process. The first one was to get information, then to eliminate subversives and sympathisers, and last was to intimidate the population. The TRC received 4,414 cases of disappearance and the whereabouts of 65% of those cases were still unknown (CVR, 2004). The forced disappearances included torture and killing applied to anyone suspected of being part of the Shining Path. The evidence of extra-judicial killings included the elimination of evidence of the disappearances.

The story of the late Angelica, mother of Atilio who was seen for the last time in 1983 tells how a group of the military broke into her house, beat her, and took her son. She said in her testimony given to the TRC, "This was the last time I saw him" (CVR, 2003). Disappearance is a military tactic to repress societies. In her daily search for truth and justice, Mendoza de Ascarza asked the soldiers where her son was. Her testimony is a declaration of obstacles she found on her daily journey. This is evidenced by the demands she presented to public institutions and different NGOs, national and international organisations, and the disregard in their responses (CVR, 2003).

The knowledge gathered from the experience of disappearances provides several narratives, depending on the angle from which they are seen. For example, the concept of disappearance for transitional justice is seen as the empirical context *versus* evidence of the case. Usually, the traditional justice system would evaluate the evidence, whereas the concept of disappearances from victims and witnesses' stories holds the real meaning of what occurred during the conflict. The truth about disappearances is compelled through an approach in experiencing the absence of the unspeakable emerging through visceral expressions. For Evans and Meza (2021), disappearance is more about the absent and the fragments that speaks and hold the pain.[9]

Los Cabitos

Los Cabitos in Ayacucho is a military base called also *cuartel 51* (base 51). In this place more than 1,000 rural people disappeared. *Los Cabitos* represented repression under the umbrella of the State of Emergency's legitimacy. Arbitrary detentions, tortures, selective freedom, disappearance, and extra-judicial killing were among the human rights abuses at *Los Cabitos*. It is noted that this obeyed a systematic practice due to the uniformity in the procedure and the high quantity of victims (CVR, 2004, p. 85). There were many military bases in Ayacucho at the time. *Los Cabitos* is the one where Angelica's son was taken after his abduction in 1983 (CVR, 2003).[10]

Los Cabitos was under the orders of Clement Noel, a graduate from the School of the Americas. In the base, crimes were committed against the *campesinos*. Noel ordered *la politica de tierra arrasada* (scorched earth policy). He did not want prisoners. The methods were the same, from arbitrary detention in military bases to torture for extrajudicial execution (CVR, 2003). The perpetrators denied any knowing to the family members in *Los Cabitos* (CVR, 2004, p. 268). The people killed were mainly thrown on *fosas* (graves) on the roads or in a place called the *infernillo*—infiernillo comes from the word hell (infierno is Spanish). Later, to remove all the evidence, the base built an oven in order to burn the people. The oven, mini sculptured in the Museo de la Memoria in Ayacucho, represents the military attempt to cover up the torture that people from Ayacucho endured. Victims were hanged by their hands to be beaten on their bodies. Electric shock tortured their intimate parts, and sometimes, their heads were put in water. Food and water were not provided adequately or at all. Military soldiers connected through the phone a family member of the torture victim, so they listened their screams. The people were abducted from their houses or by *redadas* in the streets and then tortured (see more, Los Cabitos, CVR, 2003). *Los Cabitos* represents one of the military bases which was created to carry out repression and violations in the name of security.

Community massacres

Massacres as mass killing became a strategy for both the military and the Shining Path. The experience of massacres was a critical point for displacement among the rural population. "But the war spun the world into what peasants call *chaqwa* [*ch'aqua*], Quechua for chaos and disorder. Massacres were routine by both sides. More than 600,000 were forced to flee for their lives" (Starn, 1998, p. 247). The first massacre registered was in rural Lucanamarca in 1983 executed by the Shining Path, with more than 69 people killed, 18 of them children. The majority of those living in the village were Quechua people and living in poverty (La La, 2012). Lucanamarca's massacre was pointed out by the Shining Path's leader, Abimael Guzman, in the "Interview of the Century" as an excessive action but "these things happen in war" (CVR, 2004, p. 133). With this narrative, it is seen that the head of the terrorist organisation also did not value the lives of the peasants.

In the conflict, massacres (Stern, 2005) were used widely in Ayacucho. At Shiari, the only mass killing not in Ayacucho, 62 Colons and Ashaninkas were killed with "machetes" by the Shining Path (Stern, 2005) in the jungle of Peru. Among others, in the highlands, the Uchuracay massacre of eight journalists in which those responsible remain controversial, and a prison massacre by the armed forces, with 390 dead. In the aftermath, I was a participant observer of Putis and Santo Tomás de Pata massacres' transition claim for justice. Putis was a massacre perpetrated by the military and Santo Tomás de Pata was committed mainly by the *terrucos*.

In the Putis massacre, 123 men, women and children were killed. This is another remote community in the rural area of Ayacucho situated 300 km from Lima. The area was recognised as an "active" region of Shining Path activities, hence the reason for the setting-up of a military base. The community first saw the military base was set up there to protect them from the *terrucos*. By inflicting fear from the beginning, the Shining Path threatened the people in the communities if they did not obey their commands, such as demanding food. This was some kind of imposition by both sides against the peasants in the highlands, and/or Indigenous in the jungle living in their communities.

According to El Equipo Peruano Antropologia Forense EPAF (The Peruvian Forensic Anthropology Team), the massacre started when the 123 people were called to go to the base, a "protected" part of the mountain. The military used the excuse of protection, called them in and told them to dig a "pool" for a fish farm (EPAF, n.d.). Once they finished the military members tortured them, shot them and buried them in the "pool". The military execution in Putis was the elimination of all members of the village, a familiar move for the military at that time.

Santo Tomás de Pata, the other massacre which became part of the empirical research for justice in transition, is located 60 km from the capital of Ayacucho.[11] Between the different communities within the District of Santo Tomás de Pata around 300 commuters were killed or disappeared during the conflict. When I was in this district between 2008 and 2009, it was still in extreme poverty and "full of widows and orphans", as I heard from Vargas, a community member. Finding out about their loved ones after many years, in the words of one of the relatives, meant "their souls can finally rest in peace" (International Committee of the Red Cross [ICRC], 2008).

Deprivations of freedoms

During times of conflict, the deprivations of different freedoms place everyone in a vulnerable position. The deprivations of freedoms affected at different levels their minds, bodies, homes and communities. The violent scenarios were new to the rural women and the extreme of violence put in doubt their human condition (CVR, 2004). This affected them physically, spiritually and psychologically.

Women's journeys in rural areas were compounded by different family activities such as working in their *chacras*, collecting eucalyptus for firewood or shepherding their animals. Children would help the women in these family activities when they were not going to school—the school in most cases was hours away by foot. Farming was not the main activity in Ayacucho. Their income relied on travelling between the communities and the main town, trading their local products in different market locations.

A lot of the rural areas distanced away from the main town do not have water systems and electricity infrastructure. Their activities depended on seasonal provisions and weather conditions. I mentioned earlier how oppression is manifested through lack of access to basic resources, like medical centres, schools, and police stations. The deprivation of trusted medical centres led, in many ways, to the practice of their own methods, using their traditional medicines. The lack of justice and security enabled people relied on applying communal judgement depending on the grievance.

These deprivations are indicators of disempowerment dynamics disconnecting rural women from their agency as citizens. Rural people felt deprived twice, deprived from development and deprived from peace. The majority affected were peasants and experienced deprivation of full citizenship; the Truth and Reconciliation Commission was surprised to find that 70% of the victims were not issued with a Peruvian Identification Card or DNI, as mentioned in chapter one. In this sense, deprivations during the conflict were added to the deprivations of living in poverty.

Living under chronic hardship altered their psyche corrupting rural people's mental health. Deprivation of body ownership experienced from the excessive sexual attacks as a weapon of war drove rural people to be "hidden in the cave" which led to a next deprivation, mental well-being. People living in isolated rural areas under the State of Emergency did not sleep in their homes for fear of the sudden "routines" of the *terrucos* and military. Ultimately, in fear, many did not see any other option than to be forced and displaced from their communities.

Mental health corrupted

In its aftermath, women referred to the conflict as a time of living in "craziness and hell". Theidon (2004) pointed out that woman experienced a sense of "craziness" like walking without consciousness because they could not think properly.

The women emphasised that "life was hellish" (p. 85). Their experiences during the time of violence were acknowledged as something that does not look like a normal part of this world, indeed an exposure to trauma. Caroline Yoder described a list of traumatic phases that involve threats to lives or bodies, producing terror and feelings of helplessness, overwhelming an individual's or group's ability to cope or respond to the threat, leading to a sense of loss of control, and challenging a person's or a group's sense that life is meaningful and orderly (Yoder, 2005, p. 10)

The peasants experienced traumatic stages in war. Victims-survivors were the witnesses exposed to dead and tortured bodies and other dreadful situations. Their day-to-day movements involved seeking for their loved ones in dumps of dead bodies. Consequently, as Theidon writes, they could not speak clearly because they claimed to be *andar como en un sueño* (walking like in a dream) (Theidon, 2004, p. 85). In this inhuman condition, they were constantly expressing their feelings, crying in disbelief for what they were going through. This frightened state is followed by the absence of the desire to return to their communities. Del Pino stated that:

> Las memorias del lugar, en este caso, quedan sedimentadas con las memorias del horror de la violencia; se monumentaliza el mismo paisaje geográfico como lugar de memoria y testigo de los horrores de la violencia. Asimismo, esta geografia se convierte en uno de los elementos de la fractura entre los procesos de rememorar y de olvidar (Kaufman, 1998) aquellos hechos 'duros', muy dolorosos, transformándose asi el paisaje en sitio histórico y traumático de la memoria.
>
> (2003, p. 57)

Translated by Richardson:

> The memories of the place, in this case, remain deposited with the memories of the horror of violence; the geographical landscape itself is monumentalised as a place of memory and witness to the horrors of violence. At the same time, this geography becomes one of the elements of rupture between the processes of remembering and forgetting those 'hard', extremely painful facts (Kaufman, 1998), turning the landscape itself into a historical site of memory and trauma.
>
> (2008, p. 106)

A high percentage of survivors left their *chacras* and villages and reinitiated "new" lives in other regions. Victims-survivors of any age witnessed massacres, torture, and other crimes against humanity which produced effects on their well-being. In other words, their symptoms are the content of the conceptualised trauma. Headaches and lack of sleep, among other feelings, generated their inability to concentrate on enabling the conditions to re-insert themselves into the communal life. Psychological violence was experienced by ethnic discrimination in the rural population in their everyday life (Theidon, 2004, p. 53), which also corrupted their well-being and the possibility to move on emotionally when they moved to the urban towns. Having oneself's mental health corrupted is seen as the cause for the disconnection with their *vital core.* It is a transgression of trust in their sense of reality for what they value and have a reason to value.

Another obstacle for rural women's well-being was the lack of proper documents of identification (Guillerot, 2007, p. 42), a result of their legacy of being treated by the State as second-class citizens—other authors discuss about semi-citizenship. Those obstacles contributed to stress semi-citizenship living conditions. "Semi-citizenship" is rooted in politics and inequalities (Cohen, 2009, p. 11). The TRC's realisation that a large percentage of those who died were undocumented confirmed the absence of the state institutions and their psychosocial oppression exercised through their lack of acknowledging the rural people as citizens.

Body ownership: sexual attacks

Sexual violence during repression in the Peruvian war was a tactic used to overpower the Other. Racism and sexism were the medium to dehumanise further the grievance against rural women. The TRC report evidenced that 80% of the victims of sexual violence during conflict were rural women of whom a majority had Quechua as their mother tongue (CVR, 2004). While 83% of the sexual violence is attributed to the State forces, human rights violations such as mutilations and forced domestic work is attributed to the *terrucos* (Guillerot, 2007, pp. 35–37). Another source claims the military were responsible for 98% of the sexual violations (CVR, 2004). A look at the long-term consequences of the threat of sexual violence in war for victims-survivors agency's *consiousness* and *capabilities* were ignored.

Quechua women suffered multiple violations by Shining Path members and the military, such as forced pregnancy, forced abortion, mutilation, multiple rape, and sexual slavery (Guillerot, 2006). The age of victims of sexual violence ranged from children to elderly, and in several cases the violations were in front of their husbands or/and parents. These were comitted either when the military invaded their homes looking for their partners or sons, or when women reached the military bases seeking information pertaining to the disappearance of their loved ones.

The United Nations Security Council Resolution 1820 (2008), S/RES/1820(2008): Recognizes sexual violence as a weapon and tactic of war (United Nations Security Council Resolutions on Women, Peace and Security, 2008). The consequences of rape are physical and mental, affecting the core of women's agency. The weapons behind these assaults are torture, humiliation, degradation and similar kinds of punishment. This contributes to maintain women as powerless and oppressed (Laplante, 2007, p. 315). The government's lack of listening to sexual violence claims was another obstacle to rural people's healing capabilities and left, particularly peasant, Indigenous women in the survival stage of their agency.

Stigmatisation, family rejection, feelings of guilt, domestic violence, and other reactions were the consequence of sexual violence experienced by women

(Guillerot, 2007, pp. 35–37). Sexual violence as a method of torture silence them (Henriquez Ayin, 2006, p. 72). Isolation due to fear having their husbands targeted either by the *terrucos* or the military also silenced them. Moreover, the torture for women meant also to suppress feelings when they could not speak about their abuse. According to Guillerot this suppression of feelings was due to guilt, shame and fear (Guillerot, 2007, p. 36). The consequences of being silent shows up on the openness of their physical and psychological wounds.

The harm of sexual violence was not recognised by the authorities. Whenever women empowered themselves to make a claim about rape, the authorities sought a way to silence or ignore them. This harm was an identified threat to disempower women by suppressing their agency in their courageous actions. Narda Henriquez Ayin quoted in a testimonial from the TRC report that the military punished women when they gave birth to the children of *terrucos* (Henriquez Ayin, 2006). Women were raped and punished if they attempted to make a claim for the disappearance of their family members. For example, the story of Maria Guimarita, from Chapter 1, who was tortured after lodging a complaining to the authorities about the disappearance of her husband. Rural women faced bodily vulnerability for their gender status during the conflict, whether exercising their demand for justice or not exercising it.

"Hiding in the cave"

It was frightening; during one of my research trips to Ayacucho I remember imagining myself living in the isolated rural areas during the 1980s. Spending the night in a remote community surrounded by the sounds of nature is peaceful but during the conflict the night was filled with fear. Those who did not have the option of displacement had to take the option of either hiding or defending themselves. Rural people closed their doors and walked to the mountains at night, sleeping in caves whenever they heard the *terrucos* or the military were coming.

At night the people from the villages felt threatened by home invasion, abduction or any type of violation. Sometimes they were in the caves for days and months. As quoted in testimony number 201066 cited in the CVR (TRC) report:

> In that time, because we were afraid, we had to leave [our homes] and in the mountains we had to sleep. Where we shouldn't sleep, we slept, with all my children crying, like the rain they were crying … [the local rural authorities] where they were? They also were scared, they left … everyone displaced, here and there, we had a lot of suffering.
>
> (CVR, 2004, p. 357)

The above frames the conditions that rural women had to face. This testimony represents a common experience among those living in rural Ayacucho geographically, culturally, economically, and socially far from the urban life of the capital. The *terrucos* are "abducting today!" or the military will be "raping girls tomorrow!" These were common fears and thoughts that invaded the lives of women who searched for protection in the caves, escaping with their children. Stories from surrounding communities of kidnap, rape or theft of food and abuse were faced daily.

An *Ayacuchana* commuter remembered her experience during war: "I remember my mom at night tied the dogs to the stick and they were left crying … we used to leave them for them to look after the house … and for them [Sendero] not to find us, because they used to walk on the road. Every night, for around two months, or one year, we used to go to sleep somewhere else, and not the same place, me, my mom, my older brother, my dad, and aunties" (Rodriguez Carreon, 2006, p. 67). Women were deprived of their dignity because they could be found between bushes, sleeping like animals, as testimony 301060 said "every night we were sleeping in the caves like goats or foxes" (CVR, 2004, p. 368). "They suffered with *pinsamientuwan* [worrying thoughts] that they were about to die … They suffered because they were left in the wild exposed to the harsh environment with no shelter, living as feral beasts …", said Luis Miguel, a victim-survivor (Pedersen et al., 2010, p. 287). They were deprived of their home and exposed to other threats like the weather. It was cold as low temperatures at night in the altitude of the mountains were common.

Forced community displacement

Displacement caused significant social and cultural disintegration. Rural women and children had to leave and migrate to the urban Ayacucho or to different cities. Before the conflict, the rural community cooperative network was built on trust among the *campesinos*, and then with the extreme violence these ties were broken. One of the main consequences of the broken cooperation among peasants were the need to move from their communities constantly. (Gamarra *in* Wilson, 1999, p. 148). Approximately 70% of displaced people in Peru were rural peasants from the Andes or natives from the jungle. Displacement was not a voluntary action. They were obligated to abandon their homes (ASFADEL, 2009, p. 37). Experiencing fear during displacement, they went to family and friends' houses or stayed wherever they felt safe.

Sometimes they had no time to leave with their clothes other than the ones they were wearing. They were displaced to the northern part of the country and the capital, where there is a high migration of people from Ayacucho in the south of Lima. In Lima, rural people experienced stigmatisation because of the psychological racist propaganda spread on the *identikit* image adding racism and exclusion (ASFADEL, 2009, pp. 39–41) to deepen.

Conclusion: the obstacles for human security

> "I suppose war is terrible, but isn't it picturesque?" This is the remark made by a college student as she was watching the Armistice Day parade in her city. And one had to admit that the military spectacle was thrilling.

> Ex-soldiers in full regalia, camouflaged cannons, silk flags of special units and of the nation, military bands playing stirring airs – no one could watch the procession without being stirred. But stirred with a desire to reduce armaments? Stirred with a desire to sacrifice selfish nationalism on the altars of world peace? No, emphatically no! Stirred rather with a sense of public pride in armaments, and a desire to exalt nationalism.
>
> The community celebration of the anniversary of peace served to make war attractive. The pageantry, the music, the spoils of victory, all tended to recall war and to surround it with glamor. The men who were parading knew that it was a false picture … They knew that modern war means anguish and torture of body, mind and spirit, beyond the reaches of the imagination of the cheering crowd. And yet, they were cooperating in an enterprise which made war "picturesque".
>
> (see more, Wilson, 1971, p. 18)[12]

War is raw violence. The statement above portrays the breach between the meaning of armed conflict and its antagonist representation as picturesque. War conceptualised the security narrative by rerouting the perception of fear on the processes that make conflict possible, taking us away from viewing the collective inner architecture of social dehumanisation. In Peru, the cruelty and repression of conflict contain inner qualities that affect relationships among people and communities. And yet, war, as reflected in the quote above, removes fear, absence, pain, horror—everything that war still is for the victims-survivors. In the First World War, the aversion to the rawness of war shows the lack of capability to face the inner quality of abject emotions (Biernoff, 2011). Disfigurement was everywhere on the battlefields.

The abhorrence to war's reality motivated consciousness to seek beauty on the outside. Yet, the mutilation in war is both inner and outer. The culture of aversion to seeing trauma in the face pushed professional artists to create masks for disfigured soldiers. Disfiguration meant to be judged as unfit for service, "a prosthetic mask could never replicate the warmth and expressive range of a human face, but it could hide the 'dreadful abyss' of raw being and restore a semblance of humanity" (Biernoff, 2011). Going beyond the number of deaths as the main threat of an armed conflict, the experiences narrated by survivors on their physical, psychological and community wounds shift the narrative of the real threat of war.

The cruelty of subverting entire communities to keep them powerless individually and collectively pushed rural people to change their peasantry relations and, for some, forced migration. Rural women were constantly displaced to find security that depended on their survival strategies. If those rural women were not living in poverty, they would not be treated as second-class citizens and their demands would be attended to the government.

Human insecurity lives in the rawness of war, which cruelty travels embodied as an imprint of the horrors experienced. In applying the analysis to human (in)security, the following diagram represents the experiences to which the women's agency we exposed (Figure 4.2):

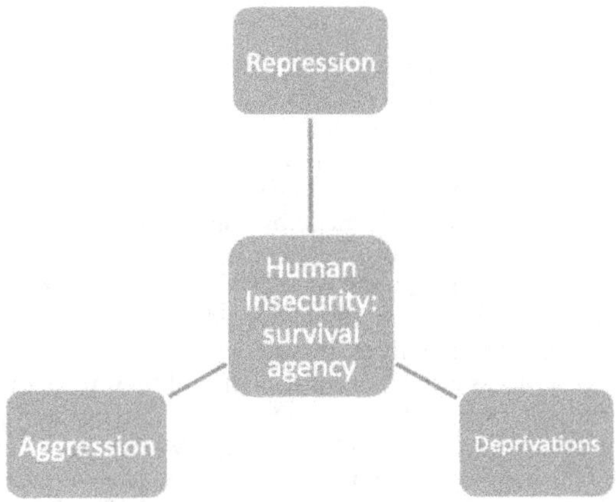

Figure 4.2 Human insecurity. ©Rodriguez Carreon, V (2014).

Figure 4.2 is a summary of how the war, through repression, aggression and deprivations in rural women, obstructed the empowerment of their agency. The terrorists and the military actions represented the cruel and barbaric violence inflicted. The terrorist discourse was under the name of freedom from want and the military under the name of freedom from fear. Humans are in uncontrolled vulnerability when the core of their agency is exposed to extreme inhuman conditions of living. And, as the experience of survivors in Peru were seen here, in some cases, they doubted their own humanity. Therefore, rural people were in the survival stage of agency.

In this chapter, the research shows how perpetrators inflicted pain on themselves and collective insecurity through fear in the population. Fragmentation spoke louder than wholeness. For the rural population, the instability broke people's trust in their communities and the system, holding back the conditions required to enhance their agency and their empowerment capabilities. Deprivations of human capabilities are one of the main factors that affect the *vital core*. However, it is affirmed that women's agency took an empowerment approach and faced the problem by looking after their endemic insecurity. This was typified by searching for food and for their loved ones (Coral Cordero cited *in* Henriquez Ayin, 2006, p. 37). Women who were transformed "positively", regardless of those deprivations, lived continuously in a disempowered environment, in which they tried to strengthen their resilience to oppression. However, such effort within any system is not sustainable. War dynamics constantly obstructed rural women's capabilities rather than allowing them to be free to enhance their human development. The inner qualities of fear and insecurity in conflict experiences changed perceptions, shaping their ways of knowing and valuing life. Attachment to inhuman experiences affected rural people's awareness. Trauma changed how some peasants interacted with their environment, separating them from their inner connection with their *vital core*, relationships with their communities, and their sacred union with the Earth.

Notes

1 Here, I refer to the word marginal with caution because they are not marginal in their identity. The literature found in poverty have a hierarchical narrative to locate identity in relation to those in position of dominating power within structure. Yet, as I mentioned in the previous chapter rural people do not identify themselves as "marginal", therefore, I can only express their identity as how they see themselves.
2 This quote comes from the analysis made by Jon Sumida of Carl von Clausewitz's on War http://www.clausewitz.com/readings/Sumida/SumidaJMH1.htm.
3 See more, Paulo Freire context of transition (1973) on critically transitive consciousness.
4 "Natural knowledge" of humans is their everyday interaction and experiences with mechanisms learnt in their daily lives. In other words, the empirical knowledge.
5 Named also "School of Assassins" as consequence of their reputation.
6 Previously watched in "School Of The Americas"—(School For Training Assassins). http://www.youtube.com/watch?v=V4eLYXJIZfg. Minute 9:29 (The case for closing the school: 17). It is not available now. Nevertheless, you can watch alternatively, the talk by Professor Lesley Gill speaking about her book The School of the Americas: Military Training and Political Violence in the Americas.
7 Who is the other? Amartya Sen in his book "identity and Violence" appoints how for example, in the case of India, through "political instigators" urged killings turning entire communities into "thugs" killing themselves. Hindus against Muslims, there was the only identity to see, nothing else (Sen, 2007 p. 172). The "other" is then, someone who cannot be seen as having the same identity, without empathy for his/her human race.
8 The SOA is known now by the name of Western Hemisphere Institute for Security Cooperation WHINSEC. This school produced several military people charged with human rights violations in Latin America. Since then, many Latin Americans have abolished to send personnel to the school. However, Peru is not in the list of those who do not send anymore soldiers to the School. http://www.soaw.org/about-the-soawhinsec/history/114-army-schools-new-name-wont-disguise-atrocities.
9 See more, in the recent project of State of Disappearance by Evans & Meza (2021). https://www.chantal-meza.com/stateofdisappearance.
10 See more, Desaparicion forzada de las personas por agentes del Estado. http://cverdad.org.pe/ifinal/pdf/TOMO%20VI/SECCION%20CUARTA-Crimenes%20y%20violaciones%20DDHH/FINAL-AGOSTO/1.2.%20DESAPARICIN%20FORZADA.pdf
11 *Santo Tomás de Pata* is part of another State, Huancavelica, which registered the majority of people living in poverty. Because Huancavelica's main town is 160 km from *Santo Tomás de Pata* the economical and social relations are closer to Huamanga, the main town of Ayacucho.
12 See more, Wilson, R. (Ed.). (1971). *Military Training.* Garland Publishing. Wilson's new edition with new introduction is from Tucker, S. (1929) *So this is War!* "Picturesque War" was published first *in* Editorial November, 1928. *International Journal of Religious Education.*

References

Aguilar, P., & Kovras, I. (2019). Explaining disappearances as a tool of political terror. *International Political Science Review*, *40*(3), 437–452. https://doi.org/10.1177/0192512118764410

ASFADEL Asociación de familias desplazadas en Lima (2009). *Atención y preparación para los desplazados en Lima por el proceso de violencia 1980 – 2000* [Care and preparation for those displaced in Lima by the process of violence 1980 – 2000].Coordinadora Nacional de Derechos Humanos.

Biernoff, S. (2011). The rhetoric of disfigurement in First World War Britain. *Social History of Medicine*, *24*(3), 666–685. https://doi.org/10.1093/shm/hkq095

Caretas (1983). *Informe especial: Estos son los comandos del ejército* [Special report: These are the army commandos]. Lima. No 730.

Cohen, E. F. (2009). *Semi-citizenship in democratic politics*. Cambridge University Press.

Coral Cordero, I. (1998). Women in war: Impact and responses. In S. J. Stern. (Ed.), *Shining and other paths: War and society in Peru, 1980–1995*. Duke University Press.

Comisión de Entrega de la Comisión de la Verdad y Reconciliación. (2004). *Hatun willakuy* [The great story]: Versión abreviada del Informe Final de la Comisión de la Verdad y Reconciliación [Abbreviated Version of the Final Report of the Truth and Reconciliation Commission]. Navarrete.

CVR Comisión de la Verdad y Reconciliación. (2003). *Desaparición Los Cabitos* [*Los Cabitos* dissapearance]. https://www.cverdad.org.pe/ifinal/

Davenport, C. (2007). *State repression and the domestic democratic peace*. Cambridge University Press.

Degregori, C. I. (1998). Harvesting storms: Peasant rondas and the defeat of Sendero Luminoso in Ayacucho. In S. J. Stern (Ed.), *Shining and other paths: War and society in Peru, 1980–1995*. Duke University Press.

Del Pino, P. (1998). Family, culture and "revolution": Everyday life with Sendero Luminoso. In S. J. Stern (Ed.), *Shining and other paths: War and society in Peru, 1980–1995* (pp. 158–192). Duke University Press.

Degregori, C. I. (2003). Introducción. In Degregori, C. I. (Ed.), *Memorias de la represión: Jamás tan cerca arremetió lo lejos* [Memories of repression: Never the so close attacked so far], (pp. 15-26). IEP.

EPAF Equipo Peruano de Antropologia Forense (n.d). *Desaparición forzada en el Perú: El aporte de la investigación antropológica forense en la obtencion de la evidencia probatoria y la construcción de un paraguas humanitaria* [Forced disappearance in Peru: The contribution of forensic anthropological research in obtaining probationary evidence and the construction of a humanitarian umbrella] EPAF.

Evans, B., & Meza, C. (2021). *Conversation of the State of Disappearance Project by Brad Evans & Chantal Meza*. https://www.youtube.com/watch?v=TdS-7Vdm3QQ

Fernandez Castillo, P. (2003). Mi nombre estaba en la lista de Sendero. In B. Jorge, D. Beatriz, & P. Gonzalo, compilers. *Para no olvidar: testimonios sobre la violencia política en el Perú* [Not to forget: testimonies about political violence in Peru]. Red para el desarrollo de las Ciencias Sociales en el Perú.

Freud, S. (1989). The unconscious. In C. Cazeaux (Ed), *The continental aesthetics reader*. Routledge.

Freud, S., & Gay, P. (1995). *The Freud reader*. Vintage.

Garrison Villard, O. (1889). Universal military training: Our latest cure-all. In R. Wilson (Ed.), *Military training* (p. 1971)). Garland Publishing.

Gertz, N. (2010). Blood/lust: Freud and the trauma of killing war. *Formations*, *1*(1). http://ojs.gc.cuny.edu/index.php/formations/article/view/674

Gill, J. (2004). *The School of the Americas*. C-SPAN. https://www.c-span.org/video/?183499-1/the-school-americas

Gorriti, G. (1990). *Shining Path: A history of the millenarian war in Peru*. University of North Carolina Press.

Guillerot, J. (2007). *Para no Olvidarlas más: Mujeres y reparaciones en el Perú* [To do not forget them anymore: Women and reparations in Peru]. APRO DEH, DEMUS, PCS.

Guillerot, J. (2006). Linking gender and reparations in Peru: A failed opportunity. In R. Rubio-Marin (Ed.), *What happened to the women? Gender and reparations for human rights violations*. International Center for Transitional Justice, Advancing Transitional Justice Series. Social Science Research Council.

Handel, M. I. (1991). *Sun Tzu and Clausewitz: The art of war and the war compared*. Professional Readings in Strategy No 2. Strategic Studies Institute. United States Army War College Carlisle Barracks.

Henriquez Ayin, N. (2006). *Cuestiones de género y poder en el conflicto armado en el Perú* [Issues of gender and power in the armed conflict in Peru]. CONCYTEC.
ICRC International Committee of the Red Cross. (2008, September 1). *Peru: uncovering the truth.* https://www.icrc.org/en/doc/resources/documents/feature/2008/peru-feature-220808.htm
ICRC International Committee of the Red Cross. (2018). *Peru: Uncovering the truth.* https://www.icrc.org/en/doc/resources/documents/feature/2008/peru-feature-220808.htm
ICRC International Committee of the Red Cross. (2010). *Missing loved ones: Helping families cope with the uncertainty.* https://www.icrc.org/en/doc/resources/documents/interview/missing-interview-150210.htm
ICRC International Committee of the Red Cross. (2019). *The missing in Latin America: Families will not stop searching, nor will we stop helping.* https://www.icrc.org/en/document/missing-latin-america-families-will-not-stop-searching
Keen, D. (2012). Shame and the psychological functions of violence. In *Useful enemies: When waging wars is more important than winning them* (pp. 195–234). Yale University Press.
Kennedy, J. (1998) *Exporting democracy or terror?* National Radio Project Transcript: #37–98 S.O.A. http://radioproject.org/transcript/1998/9837.html
La Republica. (2012). El 3 de abril de 1983, *Sendero Luminoso perpetró la masacre de Lucanamarca.* [Shining Path perpetrated the Lucanamarca massacre]. http://www.larepublica.pe/03-04-2012/el-3-de-abril-de-1983-sendero-luminoso-perpetro-la-masacre-de-lucanamarca
Laplante, L. J. (2007). Women as political participants: Psychosocial postconflict recovery in Peru. *Peace and Conflict: Journal of Peace Psychology, 13*(3), 313–331.
Oliver, S. (2011). The lived experience of dehumanization or this is not my body: Autonomy, identity, and dignity. In P. Kaufmann et al. (Eds.). *Humiliation, degradation, dehumanization human dignity violated* (pp. 90–93). Springer.
Paret, P. (2005). Einstein and Freud's pamphlet why war? *Historically Speaking, 6*(6), 14–19. https://muse.jhu.edu/article/422877.
Pedersen, D., Kienzier, H., & Gamarra, J. (2010). Llaki and Ñakary: Idioms of distress and suffering among the highland Quechua in the Peruvian Andes. *Culture, Medicine and Psychiatry, 34*, 279–300, http://www.ncbi.nlm.nih.gov/pubmed/20405314
Porter, E. J. (2007). *Peace building: Women in international perspective.* Routledge.
Quigley, B. (2005) *The case for closing the school of the Americas.* http://www.soaw.org/docs/LReview-Quigley.pdf
Reyna Hidalgo, H. (2013). *Los desaparecidos en el Perú nos importan* [Those who disappeared in Peru matter to us]. June 6. Coordinadora Nacional de Derechos Humanos https://derechoshumanos.pe/los-desaparecidos-en-el-peru-nos-importan/.
Rodriguez Carreon, V. (2006). *Violação dos direitos humanos no Peru 1980–2000: o impacto nas vítimas atingidas pela violência* [Violations of human rights in Peru 1980–2000: the impact on victims of violence]. https://tede2.pucsp.br/handle/handle/17796
Rodriguez Carreon, V. (2014). Empowerment formation: women's agency for participation in decision making within the poverty and conflict context, case of rural Peru [Unpublished Doctoral thesis]. The University of Sydney.
Sen, A. (2007). *Identity and violence: The illusion of destiny.* Penguin Books.
Starn, O. (1998). Villagers at arms: War and counterrevolution in the Central-South Andes. In S. J. Stern (Ed.), *Shining and other paths: War and society in Peru,* 1980-1995 (pp. 224-257). Duke University Press.
Starn, O. Degregory, C. I. & Kirk, R. (Eds.). (2005). *The Peru reader: History, culture and politics.* Duke University Press.
Stern, S. J. (2005). The tragedy of success. In O. Starn, C. I. Degregory, & R. Kirk. (Eds.), *The Peru reader: History, culture and politics* (pp. 124–128). Duke University Press.
Suarez, E. B. (2011). *Surviving the "Sasachacuy Tiempu "(difficult times): The resilience of Quechua women in the aftermath of the Peruvian armed conflict.* http://idl-bnc.idrc.ca/dspace/bitstream/10625/48930/1/IDL-48930.pdf

Tadjbakhsh, S., & Chenoy, A. M. (2009). *Human security: Concepts and implications*. Routledge.

Tamayo, A. M. (2003) ANFASEP y la lucha por la memoria de sus desaparecidos (1983-2000). In C. I. Degregori (Ed.), *Memorias de la represión. Jamás tan cerca arremetió lo lejos* [Memories of repression: Never the so close attacked so far], (pp. 95–134). Instituto de Estudios Peruanos.

Theidon, K. (2004). *Entre prójimos: el conflicto armado interno y la política de reconciliación en el Perú*. [Between Neighbors: the Internal Armed Conflict and the Politics of Reconciliation in Peru]. Instituto de Estudios Peruanos.

United Nations Security Council Resolutions on Women, Peace and Security. (2008). *United Nations Security Council Resolution 1820*. https://www.un.org/shestandsforpeace/content/united-nations-security-council-resolution-1820-2008-sres18202008

Wilson, F. (Ed.). (1999). *Violencia y espacio social: estudio sobre conflicto y recuperación*. [Violence and social space: study on conflict and recovery]. Estudios de Desarrollo Internacional. Centro de Estudios de Desarrollo.

Wilson, R. (Ed.). (1971). *Military training*. Garland Publishing.

Yates, P. (2006). *State of fear: The truth about terrorism*. Skylight Pictures.

Yoder, C. (2005). *The little book of trauma healing: When violence strikes and community security is threatened*. Good Books.

Youngers, C. (2003). *Violencia politica y sociedad civil en el Peru: Historia de la coordinadora nacional de derechos humanos*. Instituto de Estudios Peruanos.

Part III

Analysis

Transitions of Injustice and Justice

5 In the search for truth and justice

Perceptions among Peruvians shifted when unveiling what happened to survivors' experiences—as mothers and widows—in their search for truth and justice during and after the Peruvian conflict, 1980–2000. Continuously shocked by the inhuman conditions and atrocities they witnessed, victims-survivors raised the need to investigate the truth in the aftermath formally. The opportunity to reveal their stories under public attention was, in a way, to become acknowledged as part of Peruvian history. Rural women's agencies collectively became active in seeking the truth and justice, yet part of their agency was on survival. At the same time, they worked against adversities as they encountered new roles to overcome when seeking the truth. The meaning of justice evolved during the period of transition, from conflict to post-conflict. To what extent did rural women's search for truth and justice influence the development of their agency?

Transitional Justice (TJ) mechanisms is an opportunity to bestow acknowledgement by providing an umbrella of legitimacy during the investigations to disclose victims-survivors experiences. Hayner argues that TJ evolved from the field itself. With empirical evidence, it responded to an expectation of "accountability" that somehow addresses atrocities from the past (Hayner, 2011, p. 8). Transitional justice seeks to create a wider truth through a process, for example, via truth commissions, that opens the doors, particularly for those victims-survivors who experienced atrocities, to be acknowledged within a system that ignored them in the past.

Women victims-survivors' aim in Peru has been to find the truth. As my research deepened, I found that the truth will give them a sense of peace and the possibility to continue pursuing justice. In 2009, when I had a conversation with *Sra. Angélica* Mendoza de Ascarza, the late founder of ANFASEP (National Association of Relatives of the Kidnapped, Arrested, and Disappeared in Peru) in Ayacucho, she expressed that, questions without answers disturbed her thoughts and interrupted her normal daily life. Where is my son? Where are all the people that disappeared? What happened to them? When are we going to know the truth? (Mendoza, A, personal communication, 2009). These were the things she was worried about. The answers to these questions were what Angelica needed to give her a sense of peace, a sense of fairness in her life, and to pursue justice for her son's disappearance.

DOI: 10.4324/9781003450955-8

Since the inauguration of the ANFASEP museum in October 2004, I visited the Ayacucho organisation to listen to the stories of its founding members. There I met the courageous women and mothers. I found that their experiences of pain endured in war was wired into their memory. Trapped in their own emotional process when memories were recalled and reproduced in new experiences. Their thoughts were attached to the day of the disappearance, and the events they had witnessed. Uncertainty in relation to the truth, the whereabouts of their family members, caused them pain. A crying rural woman said: "Please, I just want to know the truth. Is he alive or dead? My children and I feel the need to know the truth". This kind of statement reflects the feeling of anguish that women were left with (Comisión de Entrega de la Comisión de la Verdad y Reconciliación [CVR, 2004], p. 360). In ANFASEP women's *survival* and *active agency* were manifested through feeling, perceiving, assimilating, and communicating their experiences.

Uncovering the truth was the principal motive in creating an "independent" commission that became the central part of transitional justice in Peru. The process provoked a period of transformation where the consequences from years under oppression and repression came to the surface. Marginalisation had implications for women's access to justice. Deprived of truth and justice, and confronted by trauma and pain when reliving memories, they struggle for inner transformation. The TRC's purpose was to find the context that led to the injustices perpetrated against victims-survivors to prevent them from occurring again. It is noted that the essential call was for "never again!" and the commission's aim was to prevent and restore society (Rotberg & Thompson, 2000, p. 3). Truth commissions—such as *La Comisión de la Verdad y Reconciliación* CVR (The Truth and Reconciliation Commission TRC) in Peru—function as the heart of transitional justice mechanisms in their attempt to restore and build peace by finding the truth for victims-survivors like the late Mendoza de Arcaza.

The Peruvian transition also focused on reparations. Reparations are opportunities to be repaired in the aftermath of war, yet they are also political and legal (Rubio-Marin, 2006). Victims-survivors did not at first seek reparations. They were more focused on getting support from the NGOs to find the dead and disappeared. Peasants demanded justice for their family members, at times, not thinking about how their own lives were affected. While reparations were a source of confusion, they were also an opportunity to foresee other needs, including mental well-being; helping to pay for their children's education after they lost the capacity to provide; recovering the bodies of those who disappeared; or punishing the perpetrators. The TRC helped to meet the needs of the victims, but only as part of recommendations rather than enforcement (Guillerot, 2006).

Seeking justice covers inner and outer experiences in a period of constant transition with a sense of unending. This chapter focuses on the experience of rural people whose families were able to grieve after decades, regardless of which forces killed them, and how some others continue their search for truth and justice.

Reconstructing the truth through the *warmisapa* and *ihma* eyes

Many men and women who survived the Peruvian conflict were left with considerable physical and psychological wounds, especially in Ayacucho. ANFASEP (2007) is the story of The National Association of Relatives of Kidnapped, Detained and Disappeared of Peru, founded by the *warmisapa* and *ihma*, the Quechua words for woman alone and widow respectively. My focus here is on the questions raised while reconstructing the truth step by step within the transitional justice process. I related in the previous chapter that the victims were mainly rural working men with family responsibilities. There were a large number of rural widows. It was found that these women mostly spoke Quechua, lived in poverty and were illiterate in Spanish. Rural women could not establish the proper demands needed on behalf of their family members who had disappeared (Tamayo, 2003, p. 117) due to language. The official documents were in Spanish and located in Lima, over 500km away from Ayacucho and 10 hours via bus. These rural women were stigmatised during and after the conflict as they were thought to be the mothers, wives or daughters of *terrucos*. Theidon (2004) points out that the *warmisapas* or widows (women that were alone) were victimised throughout the war based on the difficulties they confronted (p.131). Widows' legal claims were neglected and failed to be considered on their requests. The road to find the truth in the bureaucratic system that rural people went through was painful. Their agency trajectory in seeking the truth went from learning the language and filling in written forms to overcoming marginalisation.

At the ANFASEP centre, *warmisapas* and *ihmas* from the highlands met and they were speaking Quechua and some Spanish. A combination of language, dress in *polleras*, and ethnicity evoked a representation of being from rural areas. Generally, for the culture in domination, those characteristics were connoted as disempowering. This perception depended on the person's level of alienation, Spanish tendencies, and their belief that *serranos* were less knowledgeable. These labels on identities constructed over time were embedded in the collective field for those who felt superior in the culture of urban domination. Not all urban people think the same way, the identity attributed to peasants depends on a person's perception, and alienation from levels of power. Yet, how peasants felt about their own identity, depends on the unavoidable level of internalisation over centuries of oppression to their perception of *what it means to be a Quechua, Indigenous and/or peasant.* Many women perceived themselves to be dignified in their status as women, Quechua and from Ayacucho (see more, Thorp & Paredes, 2010). Here, I refer to how they identified themselves to construct a discourse against the countervailing narratives that describe them only in their victimhood. Oppression and its Othering add to the heaviness of victimhood and keep alive the narrative of rural disempowerment on peasants' ethnicity. I am not trying here to overshadow the existing vulnerability in rural women's identity regarding power. But to argue that the negative connotations that may arise when seeing, hearing, or reading about their identity and morality conceptualisation risk collectively embodying their ways of being with distorted perception. It is a hierarchically constructed narrative through the eyes of those who have used the power to define and blind a conceptualisation of how others identified themselves. During the conflict when the women came to realisation as part of re-exposure over their identity's vulnerability, the inward transformation of their agency went from *passive survival* to *active survival*.

ANFASEP's founder, Angelica, had looked for her son since 1982 until her death in 2017. Her son was taken forcibly by the military from her home in Ayacucho. She started her never ending path of truth-seeking. After several "visits" to the military base *Los Cabitos* and meeting several other women who were in the same situation, they began to exchange information on the whereabouts of their family members who had disappeared. They shared accounts of possible places for finding their bodies. "Where did you see a dead body? What was he wearing?", They shared when they met. They met in hidden secret places because meetings were prohibited due to the curfew as the region was proclaimed under the State of Emergency. These meetings became a popular way of exchanging information. The gatherings were held in Quechua—and as being an oral language there was not written exchange. Women arranged themselves by groups dividing themselves into sub-groups to check in several spots where their husbands, sons, fathers, or brothers might be found.

People were killed and thrown in caves and streets or in places called the *infiernillo*, a rubbish tip where bodies were thrown. Isabel Coral said that the women constantly exchanged information about physical characteristics and clothes that the people were wearing, and then they would go to those places and see if the bodies were of their family members (I. Coral, personal communication, 2009). "Like many original oral languages, Quechua is not linear" (Rodriguez Carreon, 2023). Their encyclopaedic memories of oral agency indicate that in their search for the truth, their agency was collective and active. Collective inner agency was actively evolving and manifesting through other ways of expression unseen to linear cultures. They were *active* and *survival agents* in response to their painful adversities encountered in the insecure environment.

Angelica with other three women formed the Committee of Relatives of the Disappeared. As they continued seeking for their relatives, they gained more members which strengthened their association's network. The association received help from those committed to offer assistance, such as a young lawyer who explained to the founder the procedures to follow for a formal request for information (A. Mendoza de Ascarza, personal communication, 2009). The lawyer helped to complete the forms in Spanish and then to complete the official papers to make the claims through the judiciary system in the capital, Lima. This associational agency was constantly ignored by the authorities in Lima. The women received several death threats from the military because of their insistence on their demands for information about the disappearances. Despite death threats and/or lack of acknowledgement of their inner and outer agency, their functioning capabilities came from receiving support from various individuals and NGOs, which strengthened the focus to remain visible in finding the truth about their relatives' disappearances.

The association played an important role for the women's agency in their empowerment formation. It provided them with a formal organisation to make their claims collectively. ANFASEP gave them a voice for their powerless *passive agency* to evolve inwardly and enabled them to establish survival mechanisms to continue with their lives. Over the course of a decade of fieldwork in rural areas (in and out between 2004 and 2014), ANFASEP in Ayacucho grew to be home to 700 *warmisapa* and *ihma* members, and to their stories filled with pain, suffering and courage to find the truth. For many who were living with painful memories their *vital core* was on survival, enough for their agency to be, and continue the inner transformation for ***empowerment***, or as I called in-power-moment.

The repressive measures of death threats from the government against peasants built more obstacles, and distanced even more, the possibility to provide rural citizens with fulfillment. As mentioned in the previous chapter, the State of Emergency for the *warmisapa* and *ihma* was a licence for the military to break into their houses during curfew. It meant violation of their property, their privacy, and their intimacy. The late Mendoza de Ascarza was called names, and her son was tortured without a mandate, which shows a violation of body ownership and dignity. She mentioned that she went to the Minister of Defence to tell him:

> Why did you order a curfew, is this the reason why the police can break into our homes and do whatever they want, killing people, and taking their goods?

For the TRC in their search for justice, stories like this one were crucial for transitional justice to account for evidence of the rule of law. Guillerot (2006) noted that in addition to the disappearances, the commission's report showed the actual struggles confronted by women which in many cases included threats, torture, rape, or arbitrary detentions (p. 140).

Around 20,000 widows and 40,000 orphans were left behind after the Peruvian conflict (Henriquez Ayin, 2006, p. 40). The widows' pain did not stop the search for their human security, freeing themselves from fear. *Mamá Angélica* (as she used to be known) said pain worked as the seed for those who survived to keep them going. It was the motivation that kept her agency in survival inwardly but active on the outer. They were focused on their motherhood and protector roles for the orphans. For those whose husband or son were disappeared, their pain was a catalyst to form an in-power-moment for seeking justice and responses to their claims.

The women's network established support among the members economically, socially and morally. Their association's concept was linked to the Latin American Federation of Associations of Detained-Disappeared (FEDEFAM), and registered as an association of families affected by violence. Later, the Committee of Relatives of the Disappeared adopted a new name, National Association of Families of the Kidnapped, Detained and Disappeared in areas under the State of Emergency in Peru. ANFASEP (n.d.) website said:

> Throughout the fight for truth and justice in ANFASEP, women often had to confront a society exclusionary and discriminatory, this despite their perseverance to find their relatives, and to achieve truth and justice.

Mendoza de Ascarza mentioned that ANFASEP's 700 members continued the journey in the search for the truth, the whereabouts of their family members and what happened with them. The context of war and poverty that surrounded them spread far geographically and culturally, wherever they were displaced. In assimilating the truth, new memories were formed and transformed the old ones. Searching for the truth in the outer world impacted the inner world of these rural women. In ANFASEP's *active agency*, memories of pain and courage were carried out when exerted.

Apperception process of the truth: forming memories

Memories were formed and transformed through the dynamics of the transitional process. Gamio Gehri (2009) argues that memory is something to recover within a complex experience for those who confronted direct violence and structural injustices (p. 107). When victims recalled their memories, they brought back their pain. The action can be necessary when this process means the reaching of justice and a return of their dignity (Porter, 2007, p. 133). Space for bringing the pain back was an opportunity for the process in transition to witness the pain and enhance healing. There was the option to maintain the memory while preventing it from poisoning the mind afterwards (Rigby, 2001, p. 76; Porter, 2007, p. 133). In the case of ANFASEP, the collection of the victims' clothes and personal items was a way to preserve their memory, and their past. A museum was found to be one way to remember those who disappeared. It also became a place of reunion to tell their stories. Moreover, ANFASEP was a place where their collective *vital core* as women could be seen as "sane" because it connected with the reality they knew, and they could hold together to sense what was happening. It was seen as a collective experience where their pain became, in a way, comprehensible. In their living through inhuman conditions, they found similarities and recognition. While it was painful, it was sharable and enabled connection.

In 2002 in a meeting for the anniversary of the Universal Declaration of Human Rights, stories were told with nostalgia about certain items that meant a lot to the women who fought all those years in ANFASEP. A cross was build marked "do not kill" which was taken to several protests to the Plaza of Ayacucho when asking for justice. There was also the *banderola* (flag). The flag was present at many meetings between ANFASEP and the military, and the threats that women received were "tattooed" on the *banderola*. They remembered tying the flag to their bodies because it became a symbol of the organisation, of their fight for justice. In the meetings, they always brought items that reminded them of their family members. The items reminded them that the experiences they lived were painful and real. One day, they were talking about having another flag, and where they could exhibit their items. They came up with the idea to display their flag on the third floor inside the house where they were running their meetings. When they proposed this idea to a German organisation, rural women were challenged to display not just one flag but to mount an exhibition with all the items that held meaning for the more than 20 years of their search for truth (ANFASEP, 2007). So, in that way they could own their storytelling and its reality. They were going to change and, as Sen argues, be the contributors to their own transformation (Sen, 1999, p. 53), by raising their own consciousness.

In the museum, which I visited at the inauguration and the following decade, there was a picture displayed of *mamá Angélica* holding a small paper. The note was the last news from her son Arquimedes. It was given to her by a soldier 15 days after her son was taken to *Los Cabitos*. She narrated her testimony at the TRC audience. On the note was written:

> *Mamá, no llores, no tengo nada para que me hagan desaparecer, voy a regresar seguro, no llores.*
>
> Mother, don't cry, I did not do anything for them to disappear me, for sure I will get back, don't cry.

Lidia Flores de Huamán, widow and founder member, together with Angelica had undertaken long fights for justice through the building of ANFASEP. A t-shirt was found on her husband's body after days of disappearance in the mountains. The body, already decomposing, showed signs of torture then eating by dogs. At that time the public prosecutor did not officially accede to recognise the body and provide a death certificate. The family members clandestinely had to pick up the remains and hide the burial of the body. That t-shirt represented the memory of the pain and the symbol of her search for justice.

Representational arts, from pictures to *retablos* visualised the memorialisation of women from the highlands. *Retablos* are artisan national heritage crafts that display what was happening to them day-to-day.

An ANFASEP member, Adelina Garcia, remarked, we had our kitchen where we provided food for the orphans, we have our cross, and the flag (ANFASEP, 2007, p. 49). It is all about telling the story in different ways, communicating, connecting with people. In a way, people can see in themselves the other person's experiences through remembering items, things they liked and stories about them.

ANFASEP was home for the widows' and mothers' members and their stories of suffering, like Angélica, Lidia and Adelina's experiences. Their stories together with the other mothers and widows captured the pain, suffering and the courage of the association's members in their survival and active journey to find peace.

Investigating the past: under the umbrella of transitional justice

According to Barrow (2013), a period of transition is a favourable circumstance for peace in places emerging from conflict, due to the need to investigate violations committed and determine accountability, under the establishment of a "transitional justice mechanism" (p. 34). Transitional justice enables a "historical" and "comparative" perception to understand the meaning of justice for the specific period (Teitel, 2000, p. 214) and an appeal for "accountability … after atrocity" (Hayner, 2011, p. 8). Transitional justice processes arise from the need to unveil atrocities committed in civil conflict, wars, or repression. In the case of Peru, the main goal was to address the human rights violations committed, such as "extrajudicial killings, disappearances, torture and arbitrary arrest and imprisonment" (Olsen et al., 2010, p. 11). Disrupted societies emerging from war and its violations demand a need for transformation after the conflict. Because of the psychosocial, socio-economic, legal, and political relationships overshadowing agency (Lambourne & Rodriguez Carreon, 2016, p. 89), several strategies are necessary for transitional processes to rebuild institutions. According to the International Center for Transitional Justice ICTJ:

> Transitional justice is a response to systematic or widespread violations of human rights. It seeks recognition for the victims and to promote possibilities for peace, reconciliation, and democracy. Transitional justice is not a special form of justice but justice adapted to societies in transforming themselves after a period of pervasive human rights abuse. In some cases, these transformations happen suddenly; in others, they may take place over many decades.
> (Olsen et al., 2010, pp. 10–11)

Transitional justice is defined by the United Nations as:

> Transitional justice is the full range of processes and mechanisms associated with a society's attempt to come to terms with a legacy of large-scale past abuses, in order to ensure accountability, serve justice and achieve reconciliation. Transitional justice processes and mechanisms are a critical component of the United Nations framework for strengthening the rule of law.
> (United Nations, 2010, p. 2)

In this sense, transitional justice emerged from the field of atrocity, from stories like Rosario, Angélica, and many others. To understand atrocities supported by rule of law, Rama Mani (2002) offers various definitions. There is a minimalist meaning and a more ambitious maximalist view. The minimalist indicates a "government" measure to assure the "protection" of its people within the law, and the maximalist precedes the minimalist view but is found to be incomplete. Mani takes Hayek's definition for the minimalist perspective, a "government … bound by rules fixed … [rules] which make it possible to foresee with fair certainty how the authority will use its coercive powers in given circumstances, and to plan one's individual affairs on the basis of that knowledge" (Mani, 2002, pp. 26–27).

For the maximalist perspective on the rule of law, she refers to the meaning supplied by the Organization for Security and Cooperation in Europe (OSCE), "justice [is] based on the recognition and full acceptance of the supreme value of the human personality and guaranteed by institutions providing a framework for its expression" (Mani, 2002, pp. 27–28). The rule of law in its minimalist meaning promotes the protection of people in conflict, and the rule of law in the broader view will foresee truth and include those in vulnerable situations.

It is important to note that from a broader point of view, institutions recognise and value diverse ways of expression to in-form (formed inwardly enabling invisible dynamics) the context. Applying the *consciousness-capabilities* language, transitional justice is calling to account that individual's *beings* and *doings* are recognised. The peasants' expressions were not supported by the way institutions framed the legal system. When the truth behind the atrocities was uncovered, it revealed that oppression was a big contributor to the injustices that women had to overcome.

Truth, justice, reconciliation, and reparation are some of the concepts that Latin America is familiar with in transitional justice. Several Latin American countries experimented in addressing those issues by building this very "specialist discipline" of transitional justice (Margarell & Filippini, 2006, p. 7). However, coming to terms with protracted armed violence like the Peruvian case, it was not about rebuilding something which existed before but becoming *conscient* from the type of relationship between the system and the people, and the absence of government institutions before the conflict. Raising consciousness in periods of transitions can be seen about illuminating the roots for connections and disconnections between Peruvians. Peruvian philosopher Gonzalo Gamio Gehri, in his work on memory said there was an expectation that justice meant relationship changes among those involved in the armed conflict. An inclusive way of achieving justice needs a new concept of ethics during the transition of governments (Gamio Gehri, 2009). Inclusive justice evolves from the victims-survivors' cries for their experiences to be acknowledged with the attempt to heal the underlying conditions of conflict. What is interesting to follow up is that the same idea of an inclusive justice can sometimes become a mechanism posing new obstacles to rural women's agency processes by compartmentalising and institutionalising lived experiences.

Essential types of justice in post-conflict societies are mutually buttressed and interdependent. Mani (2002) argues that to capture the concept of justice in societies with high levels of poverty emerging from conflict, it is necessary to consider a complete understanding of the story and the idea of justice. She named three types of post-conflict justice to consider: legal justice or the rule of law, rectificatory, and distributive justice. Despite the unclear definition of the rule of law, its addition of human rights into the legal framework in "war-torn" societies is clear (Mani, 2002).

Distributive justice is essential for agency to flourish in its empowerment formation. It includes the concepts of needs, rights and inequality (Mani, 2002), and for transitional justice to contribute to peacebuilding (Lambourne, 2009). Oppression became embedded within the system due to inequalities. If appropriate changes are not considered in the aftermath, then the treatment as second-class citizens will continue (Boesten, 2008) and hold back the empowerment of rural women. Even though there is a level of improvement in their full citizenship rights since the conflict, the capabilities to access truth and justice were still very limited. For ecosystems with unequal identity value, sensing worthiness as a citizen is a process that takes time, and it is reinforced through reassurance by the public systems.

Transitional justice is a phase of opportunity to build peace by transforming the patterns of the past. However, victims-survivors continue to embody oppression through the manifestation of impunity for the violence committed against them.

Inequalities embedded within the system play a dynamic part in everyday living. To pursue transformation, the roots of injustices need to be witnessed. Rama Mani in her note about distributive justice said, "redressing underlying causes of conflict located in real and perceived distributive inequalities can help to avert a relapse into conflict and prevent future conflicts, and also consolidate peace" (Mani, 2002, p. 9). Hence, transforming would mean addressing the needs resulting from victims-survivors distributive injustice, and include the in-formation from the existing ways of oppression.

The non-government organisations and transitional justice

Since during the conflict all community structures broke down, social, political, cultural, judicial and economic, there was an important need to assist victims-survivors within each area. It was also important to understand how assistance in one area depended on assistance in another area. Different non-government organisations (NGOs) worked with witnesses to understand how the disappearances occurred. Organisations accommodated with a particular need and task which included helping the victims-survivors with legal documents and procedures (legal needs), trauma (mental health), communities affected by bombs (infrastructure), economically/socially disorientated (development), assuring access to food (economic), finding the bodies (investigators), and exhuming them (forensic), collecting information (researchers), among other functions. All these broken structures are represented by areas within overall government.

Victims-survivors' needs were met in collaboration with diverse NGOs, which played an important role in the transitional justice process. The NGOs focused on distinctive areas of the human rights violations to assist victim-survivors in different ways. For example, the human rights organisation APRODEH helped initially with legal understanding. *El Equipo Peruano de Antropología Forense* EPAF (The Peruvian Forensic Anthropology Team) supported with the detection of mass graves, exhumation and identification of bodies (Cardoza, n.d.). Project Counselling Service PCS built projects for audiences regionally and nationally to advocate for women's agency. All the work done by NGOs was necessary, yet drawing attention to a particular need and compartmentalising victims-survivors' experiences diverted the possibility of a holistic understanding of the multidimensional interdependence of their needs. The NGOs offered expertise to provide support in a particular field and recruited specialised people. It was only in the aftermath of conflict that the NGOs came to realise that all needs were connected because they dealt with the same story.

During my visits to different NGOs, many of them commented that, in dealing with only one angle of the people's experiences, they confronted difficulty in addressing the victims' and survivors' needs. The conflict affected the lives of those rural people across all structures, and just considering one aspect, whether judicial, economic, or cultural, was impossible for understanding the whole story of victims-survivors. Once the NGOs realised this problem, they started to help each other by providing information and establishing connections to deliver better support. Then, the civil society organisation called *Para que no se repita* (Never again) found the need to organise the integration. The *Para que no se repita* NGO proposed a network for the various organisations available to victims-survivors. Rosa Maria Villaran shared that they were able to exchange information and build alliances to meet the victims-survivors' needs (R.M. Villaran, personal communications, 2009). The NGOs understood the interdependence of needs and recognised that failure to address one need may result in the failure to address another.

To uncover the extent of victims' needs, the process of transitional justice had to be dynamic. In investigating the past, the NGOs became a key connection for the TRC to access the victims and their stories. Accessing the truth about the victims-survivors' experiences opened other truths, and gradually, the unveiling of one part created space for the exposure of another layer of truth, until the underlying root could be witnessed. Transitional justice processes are different in every country, and unknown for the victims-survivors. For Peruvians, it was a new language to learn, and new skills to develop. Rural women did not use the term transitional justice, but their experiences produced the content. The laws of the Peruvian system were unknown and at the same time indispensable for those living in isolated communities. For example, to make formal claims for the disappearances happening in rural areas, people needed to go to the institutions in Lima, the capital. While the distance is over 500 km to the capital of Ayacucho, Huamanga, it wasn't the road trip from Lima to Huamanga that was distant; Ayacucho's cosmovision and the institutional culture presented two different worlds. Rural women as justice seekers faced an enormous disadvantage with the differences embedded within the judiciary system. The differences between oral idioms (for example: Quechua and Aymara) and literate language (Spanish) contains different ways *to be* and *to do* in the world. As mentioned, oral languages are not linear. Nevertheless, differences cannot exploit the position of people from the highlands or the jungle who, for centuries, lived under persistent oppression, and the repression received when they tried to receive justice was a galling reminder of the inhuman condition given by the system in power. So, even though rural victims-survivors were challenged in their agency capabilities to be conscious of their position as truth and justice seekers, the day-to-day systemic disempowered dynamics opened the wounds of oppression when new layers of truth were uncovered.

In the aftermath of the war in 2000, new dynamics of injustice started to emerge. In Ayacucho, a member from EPAF mentioned that the difficulty for the victims in the rural areas was to come to terms with how to access information on the whereabouts of their family members. In most of the cases the person making the enquiry knew very little about the process. EPAF members talked with various people about the same event to investigate the story. The victims went over and over through the last memory they had of their loved ones. For example, whether their family members had been abducted from their houses or workplaces, but they had no idea where were taken or by whom. When the witnesses shared their stories, the staff started to notice and get concerned about the survivors' mental health consequences (T. Molina, personal communications, 2009).

When rural people's *vital core* was compromised, they found difficulties in reaching out to the different NGOs. Part of the TJ process is to learn to uncover difficult truths, for example, how the *campesinos* became conscious of their full citizenship perception in relationship to their systems' oppression. With that, what is justice and injustice became clearer. A special commission was needed to go through the layers of truths and address the past. Hayner (2011) details the various objectives and mechanisms which may be part of the TJ process:

> A State may have a number of objectives in responding to past abuses: punish perpetrators, establish the truth, repair or address damage, pay respect to victims, and reform institutions to prevent further abuses. There may be other, larger aims as well, such as promoting national reconciliation and reducing conflict over the past, or highlighting a government's concern for human rights and thus gaining the favour of international partners. Likewise, there are a variety of mechanisms or policies to reach these objectives: holding trials; purging perpetrators from public or security posts; creating commissions of inquiry; providing individualized access to security files; providing reparations to victims; building memorials; or implementing military, police, judicial, or other reforms.
>
> (Hayner, 2011, p. 8)

Hayner acknowledges how difficult and unsuccessful were the attempts to comply with the demands of prosecuting and punishing in courts those responsible for human rights violations. The lack of prosecution due to political compromises for the TRC meant that they would be able to hear the victims-survivors search for truth and justice, yet not be able to fulfill the sense of justice. The TRC in Peru aimed to pursue truth and justice, as well as reparations and reconciliation, as the means to find solutions and then within this process to try to heal society. Healing was not a central goal for the commission. Furthermore, recommendations of reparations and moreover, reconciliation, were not likely to be reached in Peru because justice for the victims-survivors would still be a "utopia". At that time, accountability, acknowledgment and even punishment were not considered by the Peruvian government.

Collecting the truth: the role of truth commissions

The motivation behind the truth commission was to comprehend what were beyond stories, reveal who were responsible and locate the mass graves (Rotberg & Thompson, 2000, p. 7). Truth Commissions aim to explain the uncovered abuses by formally recognising them through individual and institutional accountability (Hayner, 2011). The complexity is reaching justice in the aftermath of protracted armed conflicts when the previous government was accused of crimes, and their members still exercising power embedded through their institutions.

Truth commissions emerged in 1983, starting with an Argentinian commission known at that time as the National Commission on the Disappeared, CONADEP (Hayner, 2011, p. 10). CONADEP was the first Commission to politically confront the search for who disappeared under the military regime. They operated democratically and independently to seek answers. When Raúl Alfonsin's government got elected and came into power, it found that the relationship with the military body impeded civil society's access to justice. The fear for the Commission on the new democratically elected government was that they were too vigorous in the pursuit and prosecution of the military personnel responsible for the abuses during the country's dirty war, which could deepen the breach between the military and civilians.

> Alfonsin's concern was to secure social peace, heal the wounds, and try to close this painful chapter of Argentina's recent history. Furthermore, the armed forces were capable of upsetting the transition to democratic governance if they felt unduly threatened. They had obstructed the workings of the National Commission on the Disappeared CONADEP in various ways, and members of the Commission were targeted by right-wing paramilitary groups linked to the security services. So, the new government felt constrained to operate within the parameters set by the military, and at times portrayed the human rights groups, with their persistent demands for retributive justice, as a threat to the national project of reconciliation and reconstruction.
>
> (Rigby, 2001, p. 70)

The impunity provided by giving amnesty to the military officials in Argentina was repeated in Peru. Rigby identified how the military in Argentina protected themselves. The government applied several laws following the period when crimes were committed. The Final Document on the War Against Subversion and Terrorism is a document that condemned human rights violations; however, it is stipulated in the document that middle rank military that may have "erred" were exempted because they did it "due to obedience" (obediencia debida). Finally, amnesty was given to all the military members who acted under the umbrella of repression (Rigby, 2001, pp. 70–71). Policies for impunity are opposite to human security and have an impact on the social relationality of the country. For example, the mothers in Argentina who sought answers for their sons' disappearances were restricted from accessing what happened to them, which means freedom from fear didn't disappear with the commission's mandate.

Accountability for crimes committed had more potential to be perceived as justice for victims-survivors. By contrast, reparations were seen as a "shadow for impunity". The victims-survivors' stories brought the paradoxical encounter with injustice in a TJ process. This confrontation of perceptions on the meaning of justice *versus* injustice is common among victims-survivors in the aftermath. Due to the sour history of oppression in rural Peru, there was not a clear picture of what justice looked like for victims-survivors. In other words, justice is not just a goal to obtain; its meaning needs to be perceived and embodied in people's everyday lives through interactions with their system. "Injustice is not just a *consequence* of conflict, but is also often a *symptom* and *cause* of conflict" (Mani, 2002, p. 5).

Under the circumstances described, how to establish a Commission to constitute a society with freedom from want and freedom from fear? Accountability can be compromised, if perpetrators are still part of the power structures and legally protected for their past crimes—such as the case of the military. The meaning of justice for victims-survivors was sensitive to change if perpetrators were given amnesty for the crimes committed, or if reparations attempted to make up for justice. In both cases, this would lead to distrust among the population, and internalised as injustice. Such as the Argentinian Truth Commission when amnesty wasn't taken positively by the mothers' petitions seeking the truth (see more, Gorini, 2006). In Peru, the mothers of children who were abducted and forced to commit crimes confronted difficult truths, because when the conflict ended, these children, who were now adults, returned to their villages. The lived experiences of war becomes part of the definition of what justice is for the victims-survivors.

La Comisión de la Verdad y Reconciliación CVR (The Truth and Reconciliation Commission TRC)

After 20 years of conflict, in 2001 the TRC was set up by the transitional government to start to compile stories from different regions within Peru. The conflict's first-hand witnesses were finally going to have the right to say what happened, their storytelling of their truth. After the authoritarian government of Alberto Fujimori fell, the coming transitional government with former late President, Valentín Paniagua, created *La Comisión de la Verdad*—The Truth Comission. Later the same year it was changed, by the following government of Alejandro Toledo, to *La Comisión de la Verdad y Reconciliación*—the Truth and Reconciliation Commission (Guillerot, 2006, p. 138). The commission's job was to find the truth through public hearings, communal, and personal interviews.

The main task for the Truth Commission was to set up an investigation to find the truth behind the events. It was a first step in the search for justice. In Peru, the word "justice" was not added to the Commission's name, but the mothers from ANFASEP hoped that it meant the recovery of their family members, and finding those responsible for the disappearances (CVR, 2004; ANFASEP, 2007, p. 46). The other main concern for the commission was reconciliation, it was found to be ambiguous by victims-survivors. Reconciliation is a difficult process, rural women noticed that finding the truth and justice first felt like an important step before finding reconciliation.

The TRC work was first known through the public call for hearings. A member of ANFASEP remembered being told that a Commission of Truth was developed, and everyone was looking forward to hearing about it (ANFASEP, 2007, p. 45). ANFASEP and the different NGOs played a key role for the TRC to reconstruct the way things worked during the conflict between the people and the organisations. The TRC invited victims-survivors to assist in the public hearings with the purpose to listen their experiences for the first time. Publicly communicated can be referred to as the first step to the truth. Yet, the hearings to recall memories of past events involved a painful journey for the peasants in Ayacucho. Hearing was not necessarily healing. Attending hearings meant that victims-survivors felt secure in some way to be able to tell what happened and recalled the particular day or days they were tortured, raped, witnessed massacres or communities being burnt, or when their family members disappeared. They also meant to share who was suspected of committing these crimes. Experiences of going through a truth commission hearing involved more than just telling the truth, it was a place to relive in public what happened, and it was seen as an opportunity to seek justice (CVR, 2004, p. 8). Courageously enabled by victims-survivors, memories stored in their *survival agency* exerted emotions to be witnessed. The way they were internalising the truth became articulated and was expressed in their own words.

Over 17,000 testimonies were collected by the TRC, including Rosario's case about the torture, disappearance and killing of her uncle, cousin, father and mother. For many of the victims-survivors, it was the first time they could speak about what happened. Within the TRC process there were two journeys: the inner Self-compiled a qualitative information when recalling emotional and traumatic memories, and the outer Self-expressed a linear narrative to communicate a story. Also, the hearings were space to enable rural women to talk about their expectations from the Commission.

The TRC president Salomon Lerner stated "access to justice remains, for the majority of the population, an unpaid debt of Peru's institutional history … Nevertheless, Peruvian society now knows a lot more about evils caused by the inhumane … Shining Path, and the often-brutal response of the State" (Margarell & Filippini, 2006, p. 5). The concept of acknowledgement played a key role to overcome perceptions of injustice and seed the transformation for agency. In that case "public knowledge" may to some degree provide a course for justice (Elster, 2010, p. 8), by knowing that their role as survival and active agents was recognised.

The archives created by the TRC in post-conflict Peru became the evidence of the truth that enabled the possibility to see justice, and prosecute in the future. The first papers called *fichas* were the initial interaction with their past experiences for victims-survivors. *Fichas* designed by the TRC, contained several questions to collect basic data about the people. Collecting these testimonies was when the TRC had its first "shock". As mentioned in Chapter 4 War and its rawness through cruelty and repression, nearly 70% of the victims-survivors did not possess a Peruvian ID (CVR, 2004). This information was a revelation of truth, beyond data. It revealed the absence from the State for the rural people's lives before, and during and after conflict. And showed that many other Peruvians who had died were unseen by the system and culture in power.

Collecting the truth brought light to the empirical evidence, and archives that provided implications of impunity. The TRC investigation found that death certificates were expedited years after the killings. Records of deaths after disappearances were not even registered as cases legally. For example, in Rosario's story, the death certificate document from a crime committed in 1988, was only finalised in 2002. The death certificate document found under Testimony 450075 dating November 2002, was a primary source for the TRC data collection, and a source of archival information to enrich the transitional justice literature (Olsen et al., 2010, p. 31). During the conflict grieving for rural people was life threatening, which had detrimental consequences for their *vital core*. Lacking government capabilities to safeguard the right to grieve from the perspective of human development and human security calls for implementing the identification of threats affecting the Self's inner development as a necessary measure to prevent further fragmentation of the Self. Institutions did not contribute to safeguarding people's freedom from fear. Inner safety approaches for healing were not given a priority. Collective healing needed a different quality of space, where the painful chapter is witnessed.

To have a gender section in the TRC, committed scholars conducted an exhaustive, detailed investigation to highlight its rationale. Julissa Mantilla, gender consultant for the Commission, incorporated gender analysis to include human rights violations perpetrated against women. The Peruvian TRC report incorporated a specific chapter on sexual violence. The documents gathered suggested a need for enlarging the coverage of the gender dimension due to the overwhelming amount of information. Yet, only a small part of the report received gender analysis (CVR, 2004, p. 4). On the other hand, the memorandum prepared can be found as the primary source in the Center of Information for Collective Memory and Human Rights established within the Ombudsman's office. Mantilla's report contains around 100 pages within 11 annexes of detailed information on what happened to the women during the conflict (Mantilla, 2002). Sexual violence was found to be a large component of the repression and aggression inflicted against rural women. When the TRC collected primary sources, the victims-survivors' stories became the empirical data. Their storytelling was collected, their experiences conceptualised, categorised, and archived according to the understanding of the TJ mechanism.

A meeting with the truth: *Putis* and *Santo Tomás de Pata*

Rural people were deprived from knowing the truth and accessing justice. Once the conflict ended, victims-survivors held onto their experiences. They became like the hard-drive of war, full of information. The truth is not static, it has an inner movement with outer effects. The peasants were the living evidence of the truth, witnessing and reconstructing in relationship. The Commission needed them to build history. But the truth went unanswered for the majority of victims-survivors when it came to the disappearances.

The right to mourn when a clandestine grave uncovered the bodies of their family members gave a sense of acknowledgement. As a participant observer at the funerals in post-conflict, I perceived how inwardly agency evolved through their mourning process for rural women. Some victims-survivors shared that, they were able to find some peace. Yet, when the bodies were found, they were able to confront new truths. Going back to their communities and unveiling the truth of what happened to them opened the possibility of a new consciousness about truth, memory and justice to emerge. This process responded to a need they had waited to fill for over 25 years.

In an event that represented memorialisation for Santo Tomás de Pata I met a victim-survivor, Rosalia. I shared part of her story in Chapter 3, her husband disappeared and was executed by the Shining Path because he was an authority in his community. She and another three women were seated next to her. Rosalia was from Anta, Ayacucho, where she raised her six children working, as she said, "like a man". Rosalia's husband was killed in 1984 and he was not buried at that time. She told me, the *senderistas* tortured her. They tore her clothes until she fainted. When she got up, she went to look for her husband, and found him dead in the mountains.

At the time, Rosalia said that justice for her meant support for her back when she lost her husband. She was living in poverty raising her six children. After more than two decades since those experiences, her meaning of justice is different. She told me, now she wants things for her sons and daughters if there is to be individual reparation. However, considering the number of times the system has failed them, rural people see things with disbelief and hope at the same time. The State institutions have been absent for centuries in the rural communities, and after the conflict, as Isabel Coral Cordero said to me, the rural areas were not of much interest to the government, because "peace" is again in the country (I. Coral, personal communication, 2009). In other words, for victims-survivors the government is absent again.

Unearthing part of the truth: trust for exhumations in *Putis*

Unearthing means excavating part of the truth through exhumations. Specific organisations offered to help to find out about the disappearances, the massacres and other violations. To find the whereabouts of their family members who had disappeared, which in some cases were known and untold by those living in the community, the organisations needed to connect the trust of the rural population. In the highlands, trust is something that *campesinos* lost during the conflict. People in rural areas knew about several mass graves where they suspected bodies were buried, but the commuters received constant death threats against them during the conflict. On top of that, they were not taken seriously by the authorities. Rural women, as most victims-survivors, were key source of information for any organisation or person interested in finding out the truth about mass graves. Trust needed to be cultivated and a relationship with the *campesinos* needed to be developed, with the commitment to take the matter seriously.

The forensic NGOs sought to create a project to engage the rural population in building the story around the disappeared family members. Following the TRC report, I found that psychologists, researchers, forensic scientists, and historians were involved in these projects. During the conflict, in a lot of the cases, the soldiers deployed by the government in charge ignored the search for graves. The process of exhumation by public institutions was not available to the victims-survivors. On that ground, independent organisations like EPAF (Peruvian Forensic Anthropology Team) worked to provide with part of the truth when exhumations happened. In the aftermath, forensics in Peru, committed to respond to what they considered were crimes against humanity. In their site, EPAF (n.d.) intention was to pierce the culture of silence that once guaranteed impunity for the perpetrators of those atrocities. Tanja Molina from EPAF said that a formal legal request needs to be made for them to execute their work. Before a formal demand must take place, there needs to be a witness, the person directly related to the person dead or disappeared, and evidence of the place to which the body was taken. In the conflict, some information as evidence was not known, therefore the stories were necessary, as in their narrative were elements to reconstruct what happened decades earlier. Researchers listened to the stories from different witnesses and proceeded to an investigation that could lead to an official allegation (T. Molina, personal communications, 2009).

As part of my fieldwork, I was present for part of the forensic work done by EPAF. In one of the many visits to ANFASEP, a meeting occurred between the women, forensic specialists, psychologists, journalists, and anthropologists. The meeting's aim was to communicate to victims-survivors the process for searching, finding, recognising, and claiming of bodies found in mass graves. The EPAF forensic specialists had just come back after months of searching, excavating, and finding a mass grave in the mountains. The mass graves called *fosa* were in Putis, the highlands of Ayacucho, isolated from Huamanga—part of Putis community massacre was shared in the previous Chapter 4. The reason for these women attending the meeting was to find out if the forensics had some clues about their family members. For the forensic team, the meeting was the opportunity to inform the outcome of their findings. Also, the purpose was to arrange a trip to Putis, so victims-survivors could recognise and identify the clothes and items from family members.

Psychologists from different NGOs were there to support victims-survivors to cope with the news given by EPAF. However, this support was observed as estranged, since the psychologists did not speak Quechua. The interaction which was expected to build trust between psychologist and "patients" was lost in both translation of language and culture. The psychological atmosphere of the meeting was not intimate, with a close sense of holding space. The set-up included tables and chairs although some peasants were sitting on the floor, with a table at the front for the forensic team. Many women travelled hours from the highlands and rural isolated communities to be at the meeting. They were tired. Communication design dynamics are an important part of the interaction between victims-survivors and the organisations that support them, and must be planned appropriately. Looking through the consciousness lenses, implementing ways to address rural women's needs for coping with their well-being requires a holistic approach. A cultural awareness of how peasants dealt with their own sense of Self is part of supporting them. Their idioms of distress had different ways to evoke through an oral cosmovision that contained their own challenges for victims-survivors. Oral capabilities, or opportunities that were enabled for ways of seeing their reality, must be represented from their stories perception.

This meeting with the truth over the Putis case was important for the TJ process. In this first encounter between family members and forensics, victims-survivors had the capability to attempt to reconstruct their stories. The forensics' ways of communicating facts when they found the bodies were a step forward for the investigation when they empathised with the community's pain. The forensics broke into tears while they were communicating that 123 bodies were found in the mass graves, and that of those bodies around 20 were children. The diagnosis given in ANFASEP were that people were tortured and executed. The rural women who attended the meeting also broke down and cried. Emotional release was part of constructing the truth and accessing emotions that were put aside for decades.

Many of the victims were not recognised. The journey to Putis finished with a proper funeral for the victims. This included the invitation to family members and provided them with the official death certificate. Their first legal document. They mourned for a day, and then they took the bodies to the cemetery.

I want to *pause* here, reflecting back on Putis, I remember those visits to ANFASEP. Particulalry 2008, I was pregnant, and I was not able to go to Putis' exhumation. But, as an action researcher in Huamanga, capital of Ayacucho, I was there, listening to victims-survivors' words expressing their inner movements and witnessing how they were facing their pain. As a Peruvian from rural Cusco, I sensed a connection with the space, the land. At the same time, I felt the isolation of the communities was beyond the geography and cosmovision. Being seen in the world through the lenses of oppression distanced *how* Peruvians experienced the conflict.

Feeling the truth: grief in Santo Tomás de Pata

Santo Tomás de Pata is a small rural town in the highlands, where I witnessed rural people going to be able to properly mourn and bury their loved ones. I mentioned briefly the story of this community in Chapter 3, Poverty and its forms of oppression, and its connection with underdevelopment. In this part, I narrate about grieving in transitional justice processes. Rural women came together in public for the first time in 25 years after experiencing the massacre and disappearance of their husbands. It was a ceremony conducted by Peruvian government officials who spoke mainly in Spanish, and I was sitting with the people from the communities who only spoke in Quechua. The ceremony involved several formalities and was attended by family members of those who disappeared and were killed in a massacre committed by the Shining Path. Forensics exhumed 105 bodies of mainly male *campesinos* from a mass grave in Santo Tomás de Pata, a village in Huancavelica, on the border with Ayacucho. Huancavelica concentrates the majority of people living in extreme poverty within Peru.

Santo Tomás de Pata is one of the many communities in the highlands without access to many of the basic needs. The district did not have access to water and electricity, and, schools building were incomplete and without sufficient teachers to run it. Once again, the villages around exemplify the absence of the State. Santo Tomás de Pata is still a place far away geographically, without developed infrastructure, far from the political imaginations of public offices. There are no government institutions to remind the *campesinos* from the community that their State is present, and that they are full citizens to enjoy what the State provides.

The access to justice, security and safety then was also far away. Rural women witnessed the disappearance and killing of their husbands, and were not able to mourn because they were threatened by the Shining Path. The family members knew where the bodies were located, but they never notified the authorities for fear of being killed. The bodies clandestinely buried at different places were exhumed by the forensics team, which included the Peruvian government.

Around 80% of these 105 exhumed bodies were male, which is consistent across the massacres during the war. The data below explains the different incursions between 1984 and 1995. In the District of Santo Tomás de Pata massacre: in July 1984, the military was responsible for torture and execution of 7 people; in June 1988, the Shining Path killed and destructed homes of 29; in May 1989, the military disappeared 6; in 1989, the Shining Path killed 10; in October 1989 the Shining Path tortured and executed 27; in January 1991, the Shining Path killed 18; in November 1991, the Shining Path killed 38; and October 1995, the Shining Path killed another 18 people.

The terror that this community endured in the highlands of an impoverished area of Peru shows how the Shining Path and the military destroyed the isolated villages of the country. Despite violations against human dignity committed before the killing—rape, disposition of goods, torture, forced abduction, and others—details of the type of death were not provided on the death certificates.

Having a death certificate offered the opportunity to have closure and, for some, an openness to face a new truth. As Lambourne (2009) said knowledge and acknowledgement are also part of the TJ process. In my witnessing as an action researcher, on this day, I saw the acknowledgement that Lambourne speaks about, that legitimised the death of the right to grieve. Victims-survivors and their extended families arrived at a press conference on the day before the funeral ceremony to recognise the bodies of their family members. Some were coming back for the first time since they were forced to migrate after the massacres. These official programs that included the participation of authorities contributed to the process of justice.

Once all the bodies were handed over officially to the victims' families, the coffins, all matching in white, were transferred to the main plaza, *Plaza de Armas*, to receive an "honour mass" at the Huamanga cathedral. The family members went around the plaza for a walking procession, and cries out for "truth, justice, and reparation" were voiced. The ritual of saying a final "goodbye" continued for the whole night, a Peruvian tradition, before continuing with the funeral on the following day. Candles and flowers decorated the white coffins to be mourning all night. In several cars, the coffins were taken to their communities at 5 a.m. on the next day. The final destination on the program was the community of Santo Tomás de Pata.

During the funeral, a Reparations committee's representative gave a talk announcing a water supply project for community development. After travelling by car from the capital of Ayacucho on a precarious road to the cemetery, the sense of isolation from State development and the project announcement struck a dissonance. A continuing sense of injustice is felt in several ways. Family members shared stories like, "my father had his gold teeth when I went to identify the body, and now, it is not there" (quote written in my fieldwork notes).

Reflecting on the days I accompanied the process of rural women, their lived-experience reveals an intense emotional path for victims-survivors. Agency is mute and gives a sense of passivity externally. However, internally, emotional changes were occurring, with pain, anger, relief, sadness, and confusion moving inwardly, making the transition of agency into the stage of a new conscientisation. Consciousness started the involvement in going through legal and public procedures to recognise the bodies of their family members. The victims-survivors, of whom the majority again were rural Quechua women, were given their family members' bodies in public. When the responsible officials, government members and journalists gathered in this official ceremony to acknowledge the massacre, the rural women experienced acknowledgement in their grieving process and, at the same time, reconstructed the new inner and outer architecture of their truth.

The imbalance of the truth in the aftermath: reparations

The head of the victims-survivors of Santo Tomás de Pata, Mr. Vargas, gave his testimony in which he stated: "We are still abandoned by public institutions ... we are still searching for justice ... we cannot find it ... we just find extreme poverty" (quote written in my fieldwork notes). Change has not come in over two decades after the massacre, even when reparations were claimed to be contributing to justice in the aftermath. Reparations in transitional justice requires management and resources for their execution, launch and delivery. Reparations come from repairing something that has been damaged, repairing the relationship between the State and its citizens. Mr. Vargas was referring not just to his situation but also to his community. Collective reparations proposed the development of infrastructures and services for people forced to leave their communities and also intended to remedy the State's presence. Repairing structures in the community through development programs built a sense of community, the sense of reconstructing cared for the communities. However, in 2009, the subject of reparations was seen in the government policies as political propaganda to get the trust of the communities, and overlooked when this is not needed.

Repairing the collective unconscious

On May 26, 2008, I observed the enquiries into reparations where The *Comisión Multisectorial de Alto Nivel* CMAN (High Level Multisectoral Commission) called representatives of victims-survivors and NGOs to be part of the *Registro Único de Víctimas* RUV (Unique Registry of Victims) hearing process. The *Plan Integral de Reparaciones* PIR (Comprehensive Reparations Program), *Balance y Proyecciones* (Balance and Projections) convened this national meeting at the Legislative Palace in Lima. Representatives of victims-survivors and NGOs were called to discuss reparations. The CMAN (High Level Multisectoral Commission) communicated what they had done so far and then received questions from the audience. The commission registered 2,273 communities. To be an integral transitional justice process, local, regional, and central governments are responsible for responding to the atrocities (see more, the Integral Program for Reparations in Peru, No. 28592).

The question of reparations in the period of transition is about relationships. Reparations and development programs for communities living in poverty were often confused. Reparations seen only as a development policy may overshadow the political agenda for the victims-survivors that claimed answers on the disappearances of their families. A community member from Fajardo in Ayacucho said that because of living in extreme poverty and without roads, they faced walking 10 hours to the closest province, risking their lives to the threats of violence from either the military or the senderistas. It is relevant to repeat the need for development infrastructure and services, as reparations were necessary, yet the demands for justice and well-being also needed to be addressed. Ignoring the hazards that are threatening the need to heal for well-being and rural human development might be holding back women's agency, which, in turn, such agency is to contribute to the decision-making for their community development in infrastructure and services. Rural people did not mention reparations as a motivation to foresee development. Also, reparations might be seen as a way to silence the government's moral and justice debt with the rural people.

During the Santo Tomás de Pata massacre burial ceremony, I mentioned that a reparations' authority gave a talk about development measures to bring water into the village. Water, a basic need, had never been fulfilled in the community. It was surreal listening about State development on a day that a collective funeral after decades gave people the license to grieve publicly. The dissonance needed to be witnessed. It was a liminal period where collectively, people related to face the truth. A truth that is inwardly experienced through the mirroring of their common feelings. Liminality is an opportunity to "co-create a collective shift in consciousness" to integrate divisions (Rodriguez Carreon, 2023). The moral recognition of reparation is a support for empowerment if it resembles an acknowledgement that there is damage to repair. Again, healing was not a central recognition as something that was needed as part of reparations.

The TRC made recommendations for reparations after investigating past injustices, with the aim of supporting victims. It was stated that "its mandate was to prepare proposals for repairing and dignifying victims and their relatives". The PIR undertook a scheme in response to the pain caused to victims-survivors by proposing "symbolic and economic" through "individual and collective" reparations (Guillerot, 2006, p. 139). Reparations however provided a platform for rural people to get justice capabilities (that is, to enable opportunities). But, as mentioned earlier, only if the actions in favour of reparations were genuine and not to overshadow the main interests of rural women seeking truth and justice. It is reported that the PIR included reaching those affected by displacement (ASFADEL, 2009, p. 67; Defensoría Del Pueblo, República del Perú, 2008). The plan for reparations was integral to the process which included registration of displaced people and those affected by forced disappearance (Defensoria, 2008). Reparations in the aftermath received a number of propositions for repair such as health, education, property and work (ASFADEL, 2009).

The PIR then set up a number of mechanisms to implement the reparations to those that qualified. Then, it was reported to another commission set up by the government to examine its execution. The CMAN Level approved collective reparations for 440 communities affected by the conflict of which 55 were in

Ayacucho. The type of reparation varied according to the negotiations between the community and the CMAN. However, regardless of the strategic plan to be executed, several difficulties prevented its implementation. Most of those bureaucratic procedures created complications in relation to permissions, technical issues, communications, and other matters. For example, with the obstacles in accessing medical prescriptions, a small number of victims-survivors managed to receive support from public health agencies (Defensoría del Pueblo, República del Perú, 2008). This process consequently felt for victims-survivors like a betrayal because, once again, they couldn't trust the government safeguarding their people.

Sophia Macher, a TRC member and then a CMAN representative, provided some insights into her work on reparations. She said women's participation in the reparations process was slow. Macher also advised on other aspects; for example, in reconciliation, she proposed examining the judicial system and considering a method that involves community participation, demographic characteristics, and culture. There needs to be a reform within the State in terms of mental health. Many communities are far away, and the approach towards the peasants continues to be that they need help because they live in poverty. There is a need to acknowledge that they are also victims of the conflict (See more, Comisión Multisectorial de Alto Nivel [CMAN], 2008).

Another observation I made as a participant in the public hearing was that of the five committee members assigned to evaluate reparations; only one was a female. The committee mentioned that when victims-survivors did not feel like repair, then the project failed to achieve its aims. For example, health and education were important, but if the victims did not feel these were reparations for what happened during the conflict, it was not an option for policy. Victims-survivors could get confused with projects such as the *vaso de leche* (milk cup) or others offered by the government as part of their development program in areas affected by poverty, as they considered not necessarily repairing the effects, for instance, of the exposure to extreme violence during the conflict.

Conclusion: the active agency, motivations for fairness in truth seeking

In searching for truth and justice, the seeds for rural human agency transformation emerged from being motivated to be in peace of mind, intertwining with the capability to sense fairness. When the perception of fairness changed, it altered the motivation. The basic meaning of justice as fairness (Brown, 2010; Rawls, 1971; Sen, 2010) depends on what we believe, which is informed by each person's experiences and the perception of those experiences. For example, if rural women knew what had happened to their sons and husbands, inwardly, their notion of justice would also change. Justice has a combination of factors beyond the legal implications, it is also a feeling of fairness and a sense of peace in the search of truth. The justice associated with living through poverty and conflict took different shapes and forms—the quality of injustice was more widely known for women than justice. The exertion of women's *passive agency* is an emerging power which is conceived when the need for peace of mind for justice is unveiled. The opposite of justice is injustice, a sense of unfairness that can trouble the mind. Yet, peace of mind or inner peace can include what is troubling because, in peace, there is no fragmentation. A troubled mind and injustice can become obstacles, and paradoxically, inner peace can be the space to hold those forces that plant the seed for the desired justice.

The process of seeking peace by desiring to know the truth and sense of fairness explained how the rural women from ANFASEP sought justice when they searched for bodies, recognised clothes, grieved their families, and dealt with reparations. As a result, peace with justice through a truth-seeking journey can be understood as empowerment, forming and transforming the stages of agency inwardly.

In investigating the past, truth commissions are the core of transitional justice. The commission's responsibility includes to unveil the roots of the armed conflict. In "excavating" the past, rural people found many layers of truth, including impunity for previous injustices. But there is a hidden social and cultural unconscious dimension that, in the process of "unearthing", symptoms of the collective start to become visible. In Peru, the collective unconscious affected rural people's experiences of the oppression from living in poverty, repression from living in war, and injustices in their truth-seeking journey. When storytelling became essential to the reconstruction of events, it was an opportunity to build and create a new narrative. Rural women's *passive* and *survival agency* was exerted so that their perception of justice underwent an inner change. The emerging truth is that the system obstructed the *being* and *doings* of people as agents of change. The inhuman conditions (explored in Chapter 3) blocked the agential capabilities of the peasants, the Indigenous, and the rural people from isolated areas. Revealing the injustice through the TJ process brought awareness to the public and the system in power over the pre-existing and continuing human insecurities for rural women's inner condition.

References

ANFASEP. (n.d.). *Asociación Nacional de Familiares de Secuestrados, Detenidos y Desaparecidos del Perú* [National Association of Relatives of Kidnapped, Detained and Disappeared of Peru]. https://anfasep.org/

ANFASEP Asociación Nacional de Familiares de Secuestrados, Detenidos y Desaparecidos del Perú (2007). *¿Hasta cuándo tu silencio? Testimonios de dolor y coraje* [How long will your silence be? Testimonies of pain and courage]. Ayacucho. ISBN: 978-603-45123-0-6

ASFADEL Asociación de familias desplazadas en Lima (2009). *Atención y preparación para los desplazados en Lima por el proceso de violencia 1980–2000* [Attention and preparation for those displaced in Lima by the process of violence 1980–2000]. Coordinadora Nacional de Derechos Humanos.

Barrow, A. (2013). Women, peace and security: Mainstreaming gender in transitional justice processes. In Yarwood, L. (Ed.). *Women and transitional justice: The experience of women as participants*. Routledge.

Boesten, J. (2008). Marrying your rapist: Domesticated war crimes in Peru. In Pankhurst, D. (Ed.). *Gendered peace: Women's struggles for post war justice and reconciliation*. Routledge.

Brown, C. (2010, September 28). *On Amartya Sen and "The Idea of Justice"* (Vol. 24.3). Carnegie Council, for Ethics in International Affairs http://www.carnegiecouncil.org/publications/journal/24_3/review_essays/001.html

Cardoza, C. R. (Ed.). (n.d). *Desaparición forzada en el Perú: el aporte de la investigación antropológica forense en la obtención de la evidencia probatoria y la construcción de un paraguas humanitaria* [Forced disappearance in Peru: The contribution of forensic anthropological research in obtaining probationary evidence and the construction of a humanitarian umbrella].Coordinadora Nacional de Derechos Humanos & EPAF Equipo Peruano de Antropología Forense.

Comisión de Entrega de la Comisión de la Verdad y Reconciliación. (2004). *Hatun willakuy* [The great story]: Versión abreviada del Informe Final de la Comisión de la Verdad y Reconciliación [Abbreviated Version of the Final Report of the Truth and Reconciliation Commission]. Navarrete.

Comisión Multisectorial de Alto Nivel CMAN (High Level Multisectoral Commission). (May 26, 2008). *Plan Integral de Reparaciones PIR* [Comprehensive Repair Plan]. Legislative Palace in Lima.

Defensoría del Pueblo, República del Perú. (2008). *El Estado frente a las víctimas de la violencia. Hacia dónde vamos en políticas de reparación y justicia?* [The State facing the victims of violence. Where are we going in reparation and justice policies?]. Informe Defensorial N 128.

Elster, J. (2010). *Justice, truth, peace.* http://www.iger.org/wp-content/uploads/2010/05/Justice-Truth-Peace-testo-in-inglese.pdf

Gamio Gehri Gehri, G. (2009). *Tiempo de memoria: reflexiones sobre derechos humanos y justicia transicional* [The time of memory: reflections on human rights and transitional justice]. Centro de Estudios y Publicaciones.

Gorini, U. (2006). *La rebelión de las Madres: Historia de las Madres de Plaza de Mayo. Tomo I (1976-1983)* [The Rebellion of the Mothers: History of the Mothers of Plaza de Mayo. Volume I (1976-1983]. Grupo Editorial Norma.

Guillerot, J. (2006). Linking gender and reparations in Peru: A failed opportunity. In Rubio-Marin, R. *What happened to the women? Gender and reparations for human rights violations*. International Center for Transitional Justice, Advancing Transitional Justice Series. Social Science Research Council.

Hayner, P. B. (2011). *Unspeakable truths: Transitional justice and the challenge of truth commissions*. Routledge.

Henriquez Ayin, N. Z. (2006). *Cuestiones de género y poder en el conflicto armado en el Perú* [Issues of gender and power in the armed conflict in Peru]. CONCYTEC.

Lambourne, W. (2009). Transitional justice and peacebuilding after mass violence. *The International Journal of Transitional Justice*, *3*(1), 28–48. https://doi.org/10.1093/ijtj/ijn037

Lambourne, W., & Rodriguez Carreon, V. (2016). Engendering transitional justice: A transformative approach to building peace and attaining human rights for women. *Human Rights Review*, *17*(1), 71–93. https://doi.org/10.1007/s12142-015-0376-0

Mani, R. (2002). *Beyond retribution: Seeking justice in the shadows of war*. Polity Press.

Mantilla Falcón, J. (14 May, 2002). *Informe Consultoría en Género de la Comisión de la Verdad y de la Reconciliación* [Gender Consultation Report for the Truth and Reconciliation Commission]. MEMORANDUM No 208-2002-CVR/I.

Margarell, L., & Filippini, L. (2006). *The legacy of truth: Criminal justice in the Peruvian transition.* http://ictj.org/sites/default/files/ICTJ-Peru-Legacy-Truth-2006-English.pdf

Olsen, T. D., Payne, L. A., & Reiter, A. G. (2010). The justice balance: When transitional justice improves human rights and democracy. *Human Rights Quarterly*, *32*(4), 980–1007. https://doi.org/10.1353/hrq.2010.0021

Porter, E. J. (2007). *Peace building: Women in international perspective*. Routledge.

Rawls, J. (1971). *A theory of justice*. The Belknap Press of Harvard University Press.
Rigby, A. (2001). *Justice and reconciliation after violence*. Lynne Rienner Publishers.
Rodriguez Carreon, V. (2023). Liminality: Change starts within. *Challenges*, *14*(2), 25. http://dx.doi.org/10.3390/challe14020025
Rotberg, R. I., & Thompson, D. (2000). Truth v. Justice. In Robert I. Rotberg & Dennis Thompson (Eds.), *The morality of truth commissions*. Princeton University Press.
Rubio-Marin, R. (2006). *What happened to the women? Gender and reparations for human rights violations*. International Center for Transitional Justice, Advancing Transitional Justice Series. Social Science Research Council.
Sen, A. (1999). *Development as freedom*. Oxford University Press.
Sen, A. (2010). *The idea of justice*. Penguin Books.
Tamayo, A. M. (2003). ANFASEP y la lucha por la memoria de sus desaparecidos (1983-2000). In Degregori, C. I. (Eds.). *Memorias de la represión. Jamás tan cerca arremetió lo lejos* [Memories of repression: Never the so close attacked so far], (pp. 95–134). Instituto de Estudios Peruanos.
Teitel, R. G. (2000). *Transitional justice*. Oxford University Press.
The Peruvian Team of Forensic Anthropology EPAF. (n.d.). https://epafperu.org/en/
Theidon, K. (2004). *Entre prójimos: el conflicto armado interno y la política de reconciliación en el Perú* [Between Neighbors: the Internal Armed Conflict and the Politics of Reconciliation in Peru]. Instituto de Estudios Peruanos.
Thorp, R., & Paredes, M. (2010). *Ethnicity and the persistence of inequality: The case of Peru*. Palgrave Macmillan.
United Nations. (2010). *Guidence note of the secretary-general: United Nations approach to transitional justice*. March, 2010. https://www.securitycouncilreport.org/atf/cf/%7B65BFCF9B-6D27-4E9C-8CD3-CF6E4FF96FF9%7D/TJ_Guidance_Note_March_2010FINAL.pdf

6 Participation for peace

The armed conflict exposed rural people to a higher vulnerability in their living conditions, forcing them to act on their *survival agency*. The confronting inner struggles for justice revealed to them a need to inquire about what they value and have a reason to value. As a result, different motivations empowered women's participation in rural human development. Their *active agency* efforts against adversities for their participation became the learning forces that kept them resilient in their search for peace.

While trauma is the static force in agency, motivation ignites participation. For Petri (1986) motivation happens when an inner or outer action powers a behaviour. To continue to be motivated, it is essential to keep progressing on the complex journey through what Schaub calls "integration" (Schaub, 2012), which is different from "adaptation" (Freire, 1973, pp. 3–5). Even though individuals can be motivated, at the same time, the psychological effects of trauma can be felt as an obstacle (Schaub, 2012). The symptoms of trauma from oppression and repression were entangled through experiences over generations and reproduced in the collective unconscious. Oppression, as Freire (1973) writes, makes a person "adaptive", consequently an object rather than the subject (pp. 4–5). In this sense, continuing to resist being in an "adaptive" state, rural people needed to be vigilant and reaffirm their actions through conscious participation and integration to be the subject of their agency's evolution.

When there is a conscious action of participating, it opens up the opportunity to see and evaluate the experiences that arise. The mind and body are connected by doing something they value and have a reason to value to enable *conscient agency*. For example, threats to women's agency occurred in oppressive and repressive contexts. Deprived of knowing the truth, rural people's lack of peace or *freedom from fear* became a principal need. However, for some victims-survivors, the uncertainty of what happened to their husbands and sons affected their *vital core*, their sense of living, and, therefore, their sense of love. Uncertainty put their agency into a survival state, a disconnection of their *vital core* from their agency or purpose for living. Nevertheless, in the continuous search for truth and justice, rural agency evolved through experiences that made people more aware of the situation they were confronted with. Participating in pain, for the lack of capability to know the truth, for some Quechua women, meant that pain was closer to being felt.

DOI: 10.4324/9781003450955-9

With the pain, their capacities to become aware of injustices became the primary seed for their transition from *survival agency* to *active agency*. Despite their participation and their impact on their peacebuilding initiatives, an awareness of their inner capabilities is necessary to prevent their *vital core* from being compromised by *llaki*—idiom of distress in Quechua culture.

The empirical data revealed the source of ontological knowledge to explore women's participation. What were rural women doing during the conflict and the transition after the conflict that changed their human agency's course of action and, therefore, the ramifications of their empowerment formation? I do not analyse here women as combatants—the reason is not because they do not exist; women were present as Shining Path followers and members, and this minority was primarily interested in acceding to participate in decision-making denied in other spheres. As mentioned previously, their expectations about their participation proved to be unfounded because the terrorist group also replicated a patriarchal relationship within their "party", and women were relegated to mainly domestic jobs (Coral Cordero, 1996, p. 12). Not every women were peacebuilders, I mentioned some women were also terrorists. However, in this work, when I mention "rural women", I refer to those whom I followed on their journey in their search for their family members who disappeared and were killed by either the Shining Path or the special commandos within the military. The focus here is on women as agents for peace; considering Henderson's (1994) words, women work better at peace-making (p. 1). Women are often found to participate in informal processes that are not necessarily acknowledged. Yet, they have a political impact (Meintjes et al., 2001, p. 89), which is relevant to women's empowerment in decision-making. Despite the impact, the substantial new barriers found in rural women's *active agency* for peace and development are related to recognition and visibility of their participation.

According to Ife (1995) several needs are interdependent for the participation in the development of a community, including: social, political, cultural, environmental, personal/spiritual. Many projects emphasise monetary relief and expect the unfolding of the other sectors mentioned. However, the transcendent need is highly relevant to acknowledge the importance of these five aspects. In a later work, Ife (2016) separated personal and spiritual aspects and added survival. He highlights the important value of spirituality and adds that "spiritual experience" cannot be seen to be limited to the religious arena. Contemplation, music, poetry, and experiences by the community can also contribute to development. The holistic perspective includes all the components in balance in order to have an awareness of all the "patterns" that are reinforcing the "structural oppression" (Ife, 1995). The context that Ife uses does not include post-conflict societies as such. Yet, his approach of including several areas is relevant, for his focus on people participating in their communities as pillars to inner and outer development of any community.

Rural women's roles changed individually and collectively during the war, impacting different dimensions wherever they interacted. Yet, they lack formal political participation capability because they were not acknowledged as members of a formal decision-making process. The bottleneck was the disempowering structures existing in the system, whose representatives interact daily in the invisible fabric of the social field and consciously or unconsciously contribute to densifying the existing oppression and repression. Rural women's associations worked within several dimensions in which they participated by creating survival strategies for development and building peace. The variables of empowerment in rural women are divided according to the outcome's impact obtained after the exercise of their agency during and in the aftermath of conflict.

This chapter focuses on capacities, the agency's goal and how women gained roles while attempting to build peace. They assumed positions within the community, diversifying their roles from the private sphere as providers to the family, political community representatives, and members of an association. Part of rural women's obstacles to their *conscient agency*'s capability resided in their suffering, which threatened to compromise their *survival agency* physically, psychologically, and spiritually. Therefore, I also analysed the inner struggle for transformation, a dimension of protection or resilience used for empowered participation.

The agency role: *I need to participate with my capacities*

The below quote is from the late Angélica Mendoza de Ascarza in Ayacucho, translated from Quechua:

> I could not be silent because there were so many women looking for their husbands and sons. That suffering gave me strength and courage. I wasn't scared of dying; I wasn't scared of anything. Because there weren't just a few things happening, there were too many sad things, suffering, crying and bleeding. Then, I couldn't abandon them [the women]. Even I used to not return to my home. I was surviving just drinking water in order to be with my women; they were together with me. They didn't abandon me.
>
> (A. Mendoza de Ascarza, personal communication, 2009)

She could not be silent in her motherhood role. The exercise of her capacities across spheres included a political impact on her community. Participation in decision-making relies on building up the capacities of the people living in poverty in order to impact on the power structures that exercise dominion over their lives. This is done by abolishing any possible obstacle and by reinforcing participation for political decision-making (Alkire, 2005; The World Bank, 2001). An entire group in deprivation is "expelled" from bringing potential transformation in their lives through participation (Young, 2011, p. 53). In this context, then, empowerment is intended to be the know-how for those deprived of social justice (See more, Ife, 1995). For instance, the name of the subtitle "I need to participate with my capacities" evokes the agency of an *Ayacuchana* women called Feliciana, a leader who said *si no tendria mi capacidad, no estaria liderando a las organizaciones* (If I didn't have my capacity, I wouldn't be leading organisations) (Gamarra Carrillo, 2009, p. 64). Through *active agency*, the use of the capacities for participants is the main reinforcement for empowerment.

As written in Chapter 5, the primary goal of the Association of Relatives of the Kidnapped and Detained-Disappeared ANFASEP, was to seek and find those who disappeared, and the organisation's arrangements worked around this commitment. Meetings were in the plaza, on street corners, in shops, and in other places where they could exchange information about the *infiernillo*—a name that alludes to hell, where corpses were dumped. It is noted that many rural Quechua women did not speak and write in Spanish, they exchanged this information in their oral-based technology. They formed small groups and went to the peripheries of Ayacucho to search for their loved ones. Then they would meet again and exchange ideas to restructure their plan (I. Coral, personal communication, 2009). All these interactions among the women exchanging information were made through their encyclopaedic memories.

To continue living, rural Quechua women who built resilience before the conflict accessed their inner agency resources of survival capacities. Women's *passive agency* was challenged to participate in other areas outside the household. Scholars Rigby (2001) and Gorini (2006) pointed out similar experiences of agency in Argentina exercised by The Mothers of Plaza de Mayo searching for their sons, for whom the plaza was also the public space of meetings. La Plaza de Mayo was the space to meet, protest, and spread in silence their inner struggle for justice. The differences between the "mother's groups" were in their history. The mothers of Ayacucho endured oppression, were stigmatised for their ethnicity, and were living with limited economic resources. Oppression affected the Quechua's agency dynamics when treated as marginalised and as second-class citizens by the system in domination.

After the conflict, foreign NGOs were present in Ayacucho to account for human rights abuses, pushing the national agenda to investigate. As mentioned, women were demanding the abuses during the war, with the support of several local organisations. However, after the conflict, women's interests shifted to aim towards development and continuing their fight to achieve justice (I. Coral, personal communication, 2009). Therefore, the role of collective agency from rural women needs to be recognised for their actions in peacebuilding, whether they were family providers for orphans and/or political negotiators.

***The family provider*: "I worked like a man"**

"I worked like a man" said Rosalia to me in Santo Tomás de Pata. With her story shared previously, this quote triggers a gender dimension in the hierarchical identities. She was talking about her role after her husband was killed decades ago. Rural women provided for the needs of their families and children. Man took decisions within the public arena and represented the family in political and business matters, said Isabel Coral Cordero (1996). Since colonialism, patriarchy has intensified in rural Peru, where domestic activities are for mothers and daughters. However, the word "provider" is socially read as the "breadwinner" who provides the financial resources. Other ways of providing are culturally imperceptible and contribute to the invisibilities of other roles in rural human agency.

Prior to the conflict, women were counted on for autonomy within the areas of food, health and education of their children. The health and education environment required an involvement outside the household. Women participating in the external world were confined to meetings at medical centres or schools (Coral Cordero, 1996, pp. 6–7). Their participation in those contexts was not acknowledged as exerting their agency in their lives. However, in experiencing these "domestic activities", women built knowledge for their capacities in negotiation and decision-making. Going back to Rosalia's story, for example, during the conflict, she took on both roles of parenthood and carried the sorrow of her husband's death. I added that time in my notes that she, like many Andean women, became the full participant decision-maker within the household. However, the collective culture did not see Rosalia's agency as the "breadwinner"; therefore, she rarely recognised herself as the "provider". Recognising the centuries of oppressing rural women in their ways of being in the world, to enhance rural human development, it is necessary to have a cultural collective embodiment of the freedom to acknowledge rural agency.

The political negotiator

Women's associations play a role in transforming their "traditional roles" of "motherhood" to participate in peacebuilding through social and political activities (Banerjee et al., 2010, p. 63). For rural women from ANFASEP, their motherhood identity was gradually negatively viewed, encroaching on what political agency meant socially. The role of motherhood implied using several skills that emerged in ANFASEP. They intrinsically learnt several decision-making skills in day-to-day activities within the household, unseen and categorised by the system in domination as the private sphere. Now, the mothers were provided with exposure and space into the public domain. Rural women leaders of organisations who had advanced in governing within their community during the conflict were challenged with power when dealing with the bureaucratic mechanisms of public institutions. Katrine Pedersen (2006) writes that trivial participation in the political arena is due to power being controlled generally by the aristocracy (Pedersen, 2006, p. 15), which was also observed in the Peruvian case.

We have seen in the previous chapter the different areas where women interacted publicly. They participated collaboratively in decision-making within the transitional justice process by seeking the truth, identifying bodies, making claims informally and formally, enquiring in formal letters to the authorities, protesting for their rights, acquiring information, linking random actions, suggesting changes to the curfew, and much more. Most men were not there. Women applied their oral Quechua technology to participate in progressive strategies to advance towards achieving justice. They learnt the Spanish mechanisms, and interacted with the government bodies to negotiate their rights.

Mothers were political negotiators within the local government at the informal level challenging political agency. Nevertheless, it has been pointed out that this was an opportunity to start engaging in the public arena. The limited access to the formal process motivated them to participate (Bouta et al., 2005, p. 66). "In participating in politics, whether through electoral politics, public life, or non-governmental organizations and movements, women are empowering themselves"

(Karl, 1995, p. 14). During the war, I argue that rural women leaders applied the capacities they internalised during the state of *passive agency* before the conflict—I am not referring here to the oppression but to the quality of the relationship with the land. How rural women faced challenges shows their inner condition exerted emerged from a mature resilience foundation coming from working genuinely in connetion with their Environment.

Enduring the liminal stage as inner struggle for transformation

Women's *passive agency* underwent difficulties inwardly to exert their capacities actively due to the complex embodied context of multidimensional oppression and repression. The decision-making dynamics belonged to men who built the architecture of participation, negotiations, and decisions. However, for women to survive difficult times, they produced an intrinsic change in their agency to be able to adapt to their community needs. Kieffer argues that "it's important for people to go through the process and see what you have to do to get something done in the system. It's important for them to have to feel all the agony, and all of that … growing is an awful painful process" (Kieffer, 1984, p. 27). The previous quote alludes to the rural woman's inner struggles, which brought intrinsic growth to their agency. The inner struggles are movements metamorphosed within and yet passive on the outside of their agent manifestation.

The inner struggles for participation are similar in many countries where patriarchal domination exists and conflict has been experienced. George-Williams (2005) argues that the main struggles for women in African countries, for example, are in changing their roles and then waiting for legitimacy by being submissive and holding a "subservient position". In Peru, we have seen how women changed roles during the conflict by assuming the responsibility for sustaining the family. In many post-conflict places, decision-making capabilities are lost when men return to their homes. In Peru, power structures have yet to recognise women who exerted roles in decision-making.

My emphasis on the impact of women's participation is to highlight the role of women's *active agency* and not to concentrate on the literature that portrays women as "passive victims"—such as the report about women, peace and security with case studies in Kenya, Nepal, and North-East India (George-Williams, 2005, p. 61). However, the definitions are generally applicable to countries with stigmatisation relating to power, such as Peru. "Men tell us to shut up", this is what men said to us, said women during a workshop for capacity building in the aftermath of conflict in Peru. Another woman in the group said that it is necessary to assume that machismo conditions existed because women permitted them. The dialogue was part of a workshop I attended as an action researcher on December 2009, in Huamanga, Ayacucho. Displaced rural women from different communities of Ayacucho were leaders in the urban Huamanga. The one-day workshop was part of the *Vaso de Leche* (Milk Cup) project. In this program, women have been participating since 1970 as a space to practise their social and political agency. The women's courage to take ownership of their inner experience in the submission of patriarchal dynamics was a step forward in their awareness of their empowerment process. The workshop supported women's capacities in transforming their roles. They first became aware of their former agency being seen as passive after participating in decision-making roles.

It is difficult for women who are oppressed in daily discourse to self-acknowledge that a core belief about them as "passive victims" can be shifted. In the workshop, the dynamic for these community leaders changed when their participation was acknowledged. For instance, one of the questions at the workshop was to address which part of their body they liked the most and which quality of their personality they valued. They said the exercise was difficult because they did not have time to see themselves. Then, they were told to embrace themselves, love themselves, and acknowledge what is outside and inside their agency. The exercise attempted to build self-esteem in women who complained about the rejection and insults from men. I wrote in my notes that the women recognised the dynamic as challenging. After all, women in conflict and post-conflict shared a sour context. In war, women were tortured either physically or psychologically (George-Williams, 2005, p. 62). Also, they were known for belonging to the kitchen and men outside the house. Women were to serve and men to give orders. Therefore, the discourse in the aftermath needed to emphasise women's capacities and acknowledge the difference in times of inner struggle of their learning process.

Resilience: inner growing

The ANFASEP founder told me, "they used to tell me: *Mamá*, do not abandon us, let's walk together. In some place, we will find our kids; you too; you will find your son. Do not think of abandoning us; don't get weak with your cry. Together, we will find them in the mountains, in the river. With those words, they gave me courage" (A. Mendoza de Ascarza, personal communication, 2009). She described how she gained strength in hardship. In difficult circumstances, resilience is built to grow the soul to immune the spirit. One can compare this inner change to the immune system when people get sick. The biological body creates defences for diseases by staying in pain when building immunity. Moreover, we can improve our health by eating the foods our organic body needs. Resilience can also be built by bringing peace to the mind through contemplative practices. In times that can otherwise be difficult when we confront oppression and war, mindfulness helps to grow the soul as the spirit of peace (Thich Nhat Hanh, 2003). Several people in ANFASEP built up their resilience during the conflict by participating in peaceful activities, but, in many cases, the "thoughts of uncertainty" persecuted them inwardly. Healing was not a conversation ready to occur.

People living in the highlands and the jungle were the victims of the generalised structural oppression and grotesque forms of violence. Also, peasants and Indigenous people suffered the pain of often being helpless from the violence of armed conflict and economic structures that impose cruel deprivations. Women were the victims of integrity violation by sexual abuse and rape, over their bodies (Reardon, 1993, p. 154). At the same time, ironically, they are the victims of the stigma that labelled them as "victims".

Rural women living in vulnerable contexts are exposed to the unpredictability of hazards in war. Hazards are different unexpected incidents defined as "dangerous events" with high damage impact. The vulnerable environment can affect people's livelihoods and leave the feeling of being "powerless" (Pasteur, 2011, p. 11, 16, 32). The reactions of women when they feel vulnerable can vary depending on their level of resilience, and they can become empowered to search for an outcome of the traumatic experiences. Judith Herman (1992) states that "the core experiences of psychological trauma are disempowerment and disconnection from others. Recovery, therefore, is based upon the empowerment of the survivors and the creation of new connections" (Herman, 1992, p. 133). Herman's words on empowerment invite us to search for resilience capabilities in vulnerable times.

In Chapter 3, the effects of oppression that women in the highlands of Peru faced before and during conflict are reflected in the statement made by a Sudanese refugee in Kakuma camp in Kenya. Regardless of the place, women in war confront similar hardships. "Women, especially, are full of worries and frustrations. Apart from being overworked, they face discrimination" (Vella, 2001, p. 53). In a system where the environment holds "structures of oppression" the past cannot be ignored, and in order to overcome it, the system must admit complicity in permitting such a "legacy" (Ackermann, 1996, p. 146). Moreover, the system should acknowledge that "the poor and oppressed having a voice and a participatory role, contributes to social reconstruction and reconciliation" (Laplante, 2007; Sveaass & Castillo, 2000, p. 317). Acknowledgement works as a resource. It is a hope for women's healing, which tends to grow from their stories of suffering (Ackermann, 1996, p. 146). The increasing need for having the capacity to cope constantly reinforces the support for women to participate in building resilience. Coping mechanisms, however, can overshadow their growing consciousness and distort relationality when confronting adversities. In rural communities, resilience is the ability to respond and adapt to challenges in difficult circumstances. Women's resilience embodies an internal force to keep going, yet their awareness of *how* they faced these challenges is relevant to their inner agency's conscientious evolution.

Women who experience multiple deprivations in extreme poverty and marginalisation have built capacities for being resilient in difficult times. They were psychologically and economically vulnerable to the war. Nobody could prepare them for the inhuman conditions they experienced (Project Counseling Service, n.d., pp. 9–10). In Ayacucho, several threats compromised their *vital core* through trauma: starvation, witnessing mass crime, rape, displacement, leaving their homes, losing their family members, and bombs. Poverty increased as roads were blocked, markets did not run normally, usual trading among peasants did not function, and so on. Hardship living endangered women to being exposed to traumatic experiences. Herman states that "traumatic events have primary effects not only on the psychological structures of the self but also on the systems of attachment and meaning that link individuals and community"… destroying the way victims see themselves and the world and creating different assumptions (Herman, 1992, p. 51). Not enough attention is paid to the threat of disconnecting the sense of oneself from experiencing traumatic events. A human being able to survive war, could be tolerant to hazards, yet vulnerable to illness or disconnected with their capacity to sense oneself. Elisabeth Porter states, "their [women] account of the horrors that they found is vivid, yet they also tell of women surviving trauma and finding courage to rebuild communities" (Porter, 2007, p. 18). Vulnerability context is a determinant of resilience (Suarez, 2011), yet it still can cause psychological, physical and economic fragmentations. I argue that more attention should be paid to the hazards of trauma, as they increase the disconnection with the *vital core*.

Active struggling is a *survival agency*; people are active yet do not necessarily develop consciousness of the situation. Therefore, resilience is a long-term analysis of witnessing *how* adaptive processes are developed for human development. "Positive coping strategies are an important aspect of resilience", tactics that people use based on their "skills" and "resources" to survive the hazards (Pasteur, 2011, p. 14). In Peru resilience was a process, and during this time capacities are being built to be able to cope. For example, daily women's action continued, even with the pain of their inner struggles; however, without awareness of the pain, they were at risk of reproducing their trauma and compromise their *vital core* with *llaki*.

Resilience is often seen in the literature as the capacity to survive traumatic experiences and challenges in different ways. At the same time, resilience as a coping mechanism is adapted to the function of being sane in adversity. For example, the quote from Steven Southwick et al. (2014), "Most of us think of resilience as the ability to bend but not break, bounce back, and perhaps even grow in the face of adverse life experiences". The capacity of bouncing back, according to the American Psychological Association (APA) dictionary, is when people go through a process of adaptation to adversities (American Psychological Association, 2018). However, we only see the surface, the outer, instead of the inner struggle for survival and the long-term consequences of compromising the *vital core*. The reductionist view of war to see life as either being alive or dead obstructs a robust approach to identifying threats to human security that go beyond the impetuous evaluation of an outer outcome. *Consciousness-capabilities* make the inner and outer connection a vigorous approach to a more sustainable way to see our society advance.

Inner development evolves within empowerment processes. Kieffer's research (1984) suggests four stages as part of the empowerment process: an era of entry, an era of advancement, an era of incorporation, and an era of commitment. The dynamic processes move from "socio-political illiteracy" to a mature participant whose skills progress in all four stages. His research is not based on experiences after survival; instead, his construction covered the development of empowerment as a process in transition. This definition is applicable to conflict and peacebuilding situations of rural women, who are seen as powerless and marginalised in different dimensions by the social perception of domination. However, despite the "sense of powerlessness which characterised the prior life-worlds of all participants", Kieffer said:

> These individuals also shared a common sense of "integrity" in their sensed self-identity and daily lives. Aspects of this underlying attitude include a strong sense of pride and determination, a deeply felt rootedness in community, a commitment of self-reliance, and feelings of attachment and support

> within a caring community of peers. It is only as this sense of integrity is directly violated or attacked that these individuals respond. The common or daily experience of injustice and exploitation is insufficient in its provocation. For all of the study's participants [in his research], only tangible and direct threats to individual or familiar self-interests provoke the initiation of what ultimately evolves as an empowering response.
>
> (Kieffer, 1984, p. 18)

This is the case of women in Ayacucho, including the late Mendoza de Ascarza, whose resilience has been proven by surviving conflict and building peace in several ways with the association she founded. She explains, in one of her visits to the military institution, how the military persecuted her and shot at her, but the bullets hit the walls of the San Augustin's Church. She said, "all that pain gave me more courage to keep fighting and build this association" (A. Mendoza de Ascarza, personal communication, 2009).

Her association with her empowerment process passes the four stages to which Kieffer refers. Kieffer notes that the first entry stage, characterised by physical violence against a sense of integrity, is not conscious. The second step of advancement considers the reliability of the members of the group in enablers outside the organisation. In the third stage is incorporation, which is a "self-concept, strategic ability, and critical comprehension substantially mature. Through continuing inner struggle, participants confront and learn to contend with the permanence and painfulness of structural or institutional barriers to self-determination". Furthermore, their capacities' use gives them a sense of management which is "gradually incorporated in the individuals" as a sense of being in the world. In the last stage, commitment, the inner struggle continues, and this has been acknowledged by them. Then, new skills are integrated, and they teach others how to perform and continue in their leadership roles, adapting and continuing their leadership (Kieffer, 1984, pp. 18–33). In the case of rural women the pursuit of their empowerment can be acknowledged within their limitations on the lack of capabilities in sensing the freedom to be aware.

The transition: from resilience to learning

Cirila said the school was not easy for her because she spoke only in Quechua. She recognised the difficult period of adaptation, particularly when she attended high school in her province's capital. She was accustomed to living in the *chacra*, in the mountains and recalled not liking life in the urban environment. She lost her parents in the conflict, but she remembers them with love as they encouraged her always to study and to progress because they never had the opportunity to do so. Her supportive background helped her confront the city. She liked the workshops for enhancing capacities; she was learning first to deliver what she had learnt. Her willingness and hunger for more learning motivated her to be on the empowerment path. "What gave me strength is all I have learnt", said Cirila (Gamarra Carrillo, 2009, pp. 46–62).

Cirila's learning process is full of ups and downs. It resembles the process of empowerment. Her focus to keep going overcomes potential frustrations found during the process. The adapting period starts when an unknown event is exposed, and the person develops a way to survive within the new context. She reproduced her experiences. The way she perceived the newly formed memories from those experiences depends on her embodied learning capacities. Cirila's transition had a positive outcome.

During times of conflict, rural women's agency was focused on survival, participating, and exerting their capacities without pausing to see their inner struggles. In other words, their inner movement was unseen. Nevertheless, they focused on action. Coral Cordero explains she was amazed by the number of women in the plaza. I went to sit with them she said. The main topic was human rights, the attacks and the disappearances, and the other theme was surviving … I [Coral Cordero] wanted to know the reasons for women's abrupt appearance within the public domain … They were arranging themselves exchanging photos [of the missing] … for example, they would divide themselves into four commissions; some went here and the other ones there [while she points out with her hands in two different directions], between the four they would be exchanging signs; my son has this and that, signs that enabled the four of them to look for everyone. They were illiterate [oral communication technology]. I [Coral Cordero] used to think the conversation between them was going to get out of hand. When they [the women] said, my son had a checkered shirt, it's green … It was all recalled from memory [encyclopaedic memory]; they were in every corner … the information was so rich, … the age, signs and more signs, … then some were returning already bringing information. Then, early in the morning, another expedition took place (Coral Cordero, I. personal communication, 2009). They will do it all over again; learning is participating. They were taking the initiative to know something and perform something where the process is also the goal.

Active agency: participation

I have argued that consciousness in the *conscient agency* is when our body, mind, and spirit are involved in action. Widows in different conflicts participate in several tasks, bringing hope to the chaos. Positively, women participate by integrating themselves to create a peaceful purpose for living. What is this peace? Is this participation an *active agency* or *conscient agency?* How is it possible to build even in times of much horror? Nhat Hanh explains this dynamic as one he experienced while living during the Vietnam War.

> Peace is not simply the absence of violence; it is the cultivation of understanding, insight, and compassion, combined with action. Peace is the practice of mindfulness, the practice of being aware or our thoughts, our actions, and the consequences of our actions. Mindfulness is at once simple and profound. When we are mindful and cultivate compassion in our daily lives, we diminish violence each day … the practice of mindfulness is simply to bring awareness into each moment of our lives.
>
> (Thich Nhat Hanh, 2003, p. 5)

The quote unveils the power of mindfulness through meditation. I argue, that rural people do have contemplative practices, of sensing their surroundings as a way of living, in connection to their Environment. However, gradually, this has been disrupted by the several wars their human condition endured. Many of the widows of war confronted the need for peace of mind as they struggled with uncertainty. They experienced a powerful sense of motivation that occupied their minds while participating by impacting positively with their *active agency*. When they were fully participating in building peace, they overcame to be absorbed by their inner struggle. ANFASEP members became aware of the loss in themselves, their families and communities. Their resilience was resourced in their relationship to their Environment. Their contemplative ways of being nature were while doing what they valued and had a reason to value, for example, when grazing animals. Freire argues that illiterate cultures in which the "dimensionality of time" is not within the culture don't need to connect with "temporality". They achieve a sense of their "historical nature" (Freire, 1973, p. 3) by responding to the natural evolution of their agency. In this sense, the peasants could achieve awareness of what was happening to them because they were connected with their nature. But, when conflict arrives, their "historical nature" fragmented in times of uncertainty, enabling trauma, and causing attachment to time. Disconnected from the nature-spiritual relation, they disconnected from their sense of reality, compromising their *vital core.* Thus, intrinsic development, mind, and spirit, within the human matters for the agency's well-being and effective participation. Development and human rights scholar Jim Ife (1995) wrote that personal and spiritual issues are probably the most important and are often neglected (pp. 168–175). There is an intrinsic connection between "state of mind" and what is under threat in a war that goes beyond seeing the physical life or death as the only thing that matters. Now, this does not mean that every rural person was in connection with their nature. However, war threatens the awareness of the conscious experience to be free. In other words, armed conflict obstructs the human evolution of the capability to sense oneself.

Impact in the social dimension

In previous chapters, we have seen how social community structures and tides are broken in war. Associating gave rural people a sense of community and trust among members and enabled them to belong to their identity. During the fieldwork, some rural women, regardless of being marginalised by the government, overcame threats from those in power of domination. For example, under the State of Emergency, several measures threatened Ayacucho's social fabric, as mentioned in Chapter 4. During curfews, meetings were prohibited, and arbitrary detention, home invasion, and abuse were deliberately used, threatening the inner agency capacity to sense oneself. Rural women, participating in association meetings in secret and reporting about the abuses committed during curfew, exercised their outer agency for the benefit of their families and communities.

The process of finding their husbands foresees work in the community, constantly at risk and without freedom. For example, arranging for the children to be safe while actively engaged in going to the military bases, searching "damp corpses", visiting

NGOs and meeting lawyers. At the same time, peasants living in remote areas were displaced and migrated to the provincial capital, Huamanga, to access possible claims. Living in constant emergencies under basic survival conditions and underdeveloped infrastructure, impacted rural people who were in remote areas days away by car and foot from the main town. They needed to travel, in many cases, all the way to Lima to make their claims officially, as institutions were non-existent in other cities. Their social sphere and dialogue changed, and their skills in several spheres increased. Rural women now talked about their husbands' disappearances and what they remembered of them in order to evaluate possibilities and seek solutions.

Impact of economic programs

During war, the protection of women and children are under threat from repression and agression deepening. The lack of access to basic needs for survival living under inhuman conditions risked children's physical security and, in the long term, their connection with their *vital core*. Furthermore, cruelty was reinforced by using racism, stigmatising powerlessness and creating a narrative that connoted them as "the enemy" of the State (Comisión de Entrega de la Comisión de la Verdad y Reconciliación [CVR], 2004, p. 347). However, women's resistance and survival strategies involved sacrifices to protect their children and themselves and impacted their economic sphere by designing programs in their associations.

While many of these programs originated for development and were not designed for conflict, they changed in form during conflict to fit their new needs. They established the *comedores populares* (people's refectories) to feed their children and the armed conflict orphans under secure shelter. Women's associations like ANFASEP implemented programs to support their members' truth-seeking. They needed some minimum economic capabilities to pursue justice for those living in remote communities, kilometres away from ANFASEP. Many women travelled under challenging circumstances to participate in this program. One of the driving forces enhancing agency's empowerment in this program was the known *mamá Angélica*, she said:

> there was not any place where women could eat, the women move from their *chacras* [farms] and did not have anywhere to eat, to sleep and where to take their children. I went to ask for some bones at the *plaza* and cook a soup, then, the second time they [the people that gave them some ingredients] did not want to collaborate with us … the women were crying because they did not have food for their children who were starving and crying. I organised, and I went to Lima to talk to the priest, and he was the one who bought the kitchen and pots … we did not have any money for the space, first a professor lent it to us [a space]. After that, I went to Lima, and I talked with the [Human Rights] coordinator, and she used to send me money and this is how we started the *comedor* [refectory]. First came ten children, after that twenty, then thirty.
>
> (A. Mendoza de Ascarza, personal communication, 2009)

Popular refectories or people's kitchens were programs active in urban areas before the conflict. Coral Cordero, who worked with women from Ayacucho for more than a decade, said these programs did not interest rural women before the conflict. Nevertheless, when the NGO *Centro de Promoción y Desarrollo Poblacional* CEPRODEP (Center for Promotion and Population Development) under Coral Cordero's management surveyed how many women would be interested in participating in *comedores populares* to give space to women's representation in social, political and economic dimensions during the conflict, the answer was "yes". Consequently, 1,700 popular kitchens were built, with more than 80,000 women participating to implement the program (I. Coral, personal communication, 2009).

Development with political impact

In general, any group association in war time had no access to the right of meeting. The government's politics of fear implemented under the State of emergency did not ensure people' safety in organising and participating in programs for political impact. Martha Nussbaum argues that: "Women in many nations have a nominal right of political participation without having this right in the sense of capability: for example, they may be threatened with violence should they leave the home. In short, thinking in terms of capability gives us a benchmark as we think about what it is to secure a right to someone" (Nussbaum, 2000, p. 98). If we apply Nussbaum's approach to what happens in armed conflict, it could be said that, in terms of legal capabilities, women's rural participation was prohibited. However, I argue that political impact existed informally because policies designed by rural women affected the governance of their communities.

Some women leaders who did pursue "political participation" had understood that one way to make changes for their communities and their people was by, participating in decision-making, as shown in their fight for justice during the transition period, as narrated in Chapter 5. Rural women's political participation endured marginalisation, discrimination and racism injuries reviewed in Chapter 3 that could jeopardise their sense of agency. They aimed to use their capacities and achieve the changes necessary to survive, first, from the abuses experienced during the conflict. Coral Cordero said, as part of the truth-seeking persisting in the search for justice, in post-conflict, women were more focused on the development and the consequences of war, such as mental health issues (I. Coral, personal communication, 2009). Their *active agency* had a political impact, yet it was not formally acknowledged after the conflict.

Lederach (2005) provides an example of how women in conflict committed themselves through association to work against their inner struggles and towards peace. As mentioned in Chapter 2, women started with the spaces they had already conquered—the markets. Then, they planned how those spaces would save them from violence. They met in the markets to make further arrangements and expand peace spaces. They used negotiation skills to reach agreements with different groups in various settings. This impulse that makes women start their associations towards peace and simultaneously impacts several other dimensions—like improving the economy within the market—is called by him the moral imagination (Lederach, 2005). Participation through collectively associating themselves was *active agency* to experience and create new possibilities with positive political impacts.

Conclusion: enabling women's capabilities for their empowerment

Participation enables the agency's capabilities to transition from passive to active. This intrinsic process occurs when rural women question their own yearning while thinking about protesting for the whereabouts of their family members or when they engage in finding new outcomes. Their agency transiting the liminal space is when they are becoming aware of their reality. What happened in this period I found is the, "naïve transitivity" described by Paulo Freire. He argues that in this part of the process, people tend to over-simplify issues, have "nostalgia" for their past, "underestimate" the "common" people, show a "lack of interest in investigation", expose the "fragility of argument" (Freire, 1973, p. 18). Even though all the characteristics cannot be applied to rural women's *being* and *doings* in the Ayacucho context, women confronted a transit stage in their agency whenever they constructively questioned the actions taken against them. The transitivity concept analysed by Freire is relevant here because it opened the door to enable the capabilities of awareness and the possibility of *conscient agency* to emerge. Or, in his words, it can be moved to the stage of "critical transitive consciousness", a stage where the person acknowledges that everything is possible. I reason that *active agency* transited from the *passive* and *survival agency* period, not leaving definitely any stage of the agency. *Active agency* exerted the core beliefs learnt from the experience in *passive* and *survival agency* in which women's thoughts and capacities became embodied. Participation enables people to see what motivations drive their *active agency*, opening the possibility of raising their consciousness. However, this period is critical because the active agent can adapt and attach to their new identity; without awareness, the identity fragments the participant from the role, and participation is no longer a change maker.

Participation enables people to see what motivations drive their *active agency*, opening the possibility of raising their consciousness. However, Motivations are driving forces for striving against an agency's inner struggles. In Peru, when the conflict was over, those who survived traumatic events and participated in different activities improved the development of their agency. However, as seen in the previous chapter, during transitional justice, traumatic memories were relived with the public hearings and exposed their identities to be manipulated into new narratives. In systems with structural oppression and hierarchical perception of identities, there is no freedom for agency evolution, and each shift is an inner struggle. In the cases of rural women from ANFASEP, their participation in positive outcomes in their livelihoods has been the core of their *active agency*. They were active and coping, but *active agency* does not necessarily equal *conscient agency*.

We have seen women behind the scenes, unpaid, collaborative, and unrecognised as actual peacebuilders, who were consistently excluded from the formal peace negotiation processes and public political decision-making (Porter, 2007).

A new narrative is needed; we cannot undermine women's participation in their empowerment process even though they were held back in their decision-making by the structures of domination perceived in power. In Peru's rural human development, disempowering people's empowerment roles through the different systems of oppression and repression—explained in Chapters 3 and 4—are structurally rooted in context. Consequently, the empowerment formation for participation in decision-making is historical systematised with the inner architecture of the collective. More attention should be paid when participation in new roles emerge in different dimensions as people change inwardly with new experiences.

During the war, women who experienced similar inner struggles became associated. In this context, courage became necessary to exercise agency and actively search for ways to move in spaces more freely for the things they value and have a reason to value. Freedom in their capabilities enhances empowerment opportunities for their agency to evolve. For example, Feliciana mentioned in her story the importance of developing her capacities. Her strategies for living and looking after her children while confronting periods of displacement and loss depended on her capabilities. She said: "if I didn't have my capacities, I would not be the leader of my organisations" (Gamarra Carrillo, 2009, p. 82).

The motivations for peace are found in the process itself. "Peace is more of a process than an objective. It is a fruit of new attitudes, a way of life ..." (Henderson, 1994, p. 2). In the empirical findings, ontological peace emerged from experiences ignited by the motivation to escape violence through non-violent channels. Organisations in peacebuilding tend to ignore the unseen inner qualities of seeing and sensing gained in *passive agency*. For example, in rejecting pain, relate the senses of inner struggles, like a battle happening inside sends the message to be pushed away rather than looking into, observing the pain. A way of living that could provide a tranquil state of mind and enable a system to make decisions without constraints. Yet, in the long term, if sustainable peace is sought, the pain that brings inner struggle ought to be acknowledged as part of making peace with the collective inner agency that supports the inner transition from *active* to *conscious agency*.

In analysing people's participation in decision-making, the question raised is why rural women would stop or slow down the growth of their capacities to exercise their power in the aftermath. The late Pilar Coll, former executive secretary of the National Coordinator of Human Rights in Peru (Coordinadora Nacional de los Derechos Humanos, CNDH), said that women from the highlands living in poverty and illiterate in Spanish did not have access to a development context. Women, unaware of the mechanisms in the decision-making spheres, lack resources, and development was an imaginary, non-existent experience for them ... (P. Coll, personal communication, 2009). As Coral Cordero added, women in the Andes lack of knowledge of urban mechanisms—where development was more likely to exist—were not of interest to the government. The government just wanted to end the conflict (I. Coral, personal communication, 2009). What the government saw as a conflict from the perception of Lima, was not the complex intersection of living in poverty, oppression of their connoted identities with the war. In this sense, the government failed not just to provide economic development for rural people but to cooperate in acknowledging the importance of bridging the gap in decision-making. In other words, it contributed to the narrative that control is needed for power in decision-making.

Capacities provided for women's participation in associations helped break the knowledge gaps from power dynamics. However, for those NGOs, supporting one solo aspect of women's vulnerabilities during conflict meant the inability to acknowledge other capacities or the lack of them. For example, Rosa Villaran from the civil society movement *Para que no se repita* PQNSR (Never Again Citizens Movement), said that NGOs needed to be inter-connected to be able to deliver an integrated project (R. Villaran, personal communication, 2009). During the conflict, women continued to participate in specific areas. When the NGOs worked together, they could see, for example, that participating in one role—politics—could empower them by bringing capacities from functions exercised in other activities—the household. Transferring participation among roles through raising consciousness enhances the overall agency.

The narrative of women's identity as victims was reinforced during the conflict and in the aftermath, and their identity as decision-makers in their own lives has not been recognised. It can be argued that phenomenologically, societies have, in their consciousness, embodied perceptions of identities, where some are connoted as powerless and unable to make decisions, and yet, this is relevant to the time and place in their context. Nevertheless, rural women in Peru, intertwined their functionings—their *being* and their *doings*—and exerted their agency as family providers, political negotiators, and other roles. Their participation had a social, economic and political impact. Several skills channelled a peacebuilding path, for example: compassion (when they fed the orphans), friendship (when they met to search for those who had disappeared), and/or coherence (when they figured out to meet rearrange the community life in different areas). Even when this included unclear moments, and there is stress caused for the poor and marginalised (Henkel & Stirrat, 2001, p. 171) rural human development is happening. Empowerment is when the process involves humans growing and becoming resilient, and understanding its process involves bringing awareness to the capability of holding the inner space to feel the contradictions, for example, the victim and the decision-maker. Then, rural women can move from experiencing the inner struggle as an outer struggle to evolve into their consciousness. Transitions enabling capabilities for *conscient agency* in decision-making exist. However, systems holding rural human development back from exercising their empowerment also need consciousness to close the power gap in participation.

References

Ackermann, D. M. (1996). The alchemy of risk, struggle, and hope. In M. J. Mananzan, M. A. Oduyoye, E. Tamez, S. Clarkson, M. Grey, & R. Letty (Eds.), *Women resisting violence: Spirituality for life* (pp. 147–160). Orbis Books.

Alkire, S. (2005). *Subjective quantitative studies of human agency*. Oxford Poverty and Human Development Initiative OPHI. University of Oxford.

American Psychological Association. (2018). Resilience. *APA Dictionary of Psychology*. https://dictionary.apa.org/resilience

ANFASEP. (n.d.). *Asociación Nacional de Familiares de Secuestrados, Detenidos y Desaparecidos del Perú* [National Association of Relatives of Kidnapped, Detained and Disappeared of Peru]. https://anfasep.org/

Banerjee, P., Poutiainen, P., Ishita, D., Dunghana, S., Kioko, W., Mattila, P., & Muhindi, M. (2010). *Women, peace and security: Implementation of UN Security Council Resolution 1325 in the context of Finnish development policy, with case studies from Kenya, Nepal and North-East India*. Finland Ministry for Foreign Affairs.

Bouta, T., Frerks, G., & Bannon, I. (2005). *Gender, conflict and development*. The World Bank.

Coral Cordero, I. (1996). *Las mujeres en la guerra: impacto y respuestas* [Women in war: impact and responses]. CEPRODEP Centro de promoción y desarrollo poblacional.

Comisión de Entrega de la Comisión de la Verdad y Reconciliación. (2004). *Hatun willakuy* [The great story]: Versión abreviada del Informe Final de la Comisión de la Verdad y Reconciliación [Abbreviated Version of the Final Report of the Truth and Reconciliation Commission]. Navarrete.

Freire, P. (1973). *Education for critical consciousness*. Sheed and Ward.

Gamarra Carrillo, J. (Ed.). (2009). Reconocimiento y ejercicio de derechos de mujeres Ayacuchanas afectadas por el conflicto armado interno [Weaving lives: Story of six women from Ayacucho]. In Proyecto Comisión Europea. *Tejiendo vidas: Historia de seis mujeres Ayacuchanas*. Biblioteca Nacional del Perú.

George-Williams, D. O. (2005). Changing gender roles in conflict and post-conflict situations in Africa. In D. Rodriguez & E. Natukunda-Togboa, *Gender and peacebuilding in Africa* (pp. 59–72). University for Peace.

Gorini, U. (2006). *La rebelión de las Madres: Historia de las Madres de Plaza de Mayo. Tomo I (1976-1983)* [The Rebellion of the Mothers: History of the Mothers of Plaza de Mayo Volume I (1976-1983)]. Grupo Editorial Norma.

Hanh, T. N. (2003). *Ending violence in yourself, your family, your community, and the world: Creating peace*. Free Press.

Heiko, H., & Roderick, S. (2001) Participation as spiritual duty: Empowerment as secular subjection. In U. Cooke & B. Cooke (Eds.) *Participation: The new tyranny?* Zed Books.

Henderson, M. (1994). *All her paths are peace: Women pioneers in peacemaking*. Kumarian Press.

Herman, J. L. (1992). *Trauma and recovery: The aftermath of violence–from domestic abuse to political terror*. Basic Books.

Ife, J. (1995). *Community development: Creating community alternatives – vision, analysis and practice*. Longman Australia.

Ife, J. (2016). *Community development in an uncertain world: Vision, analysis and practice* (Second ed.). Cambridge University Press.

Karl, M. (1995). *Women and empowerment: Participation and decision making*. Zed Books.

Kieffer, C. H. (1984). Citizen empowerment: A development perspective. In J. Rappaport, C. Swift, & R. Hess (Eds.), *Studies in empowerment: Step toward understanding and action* (pp. 9–36). The Haworth Press.

Laplante, L. J. (2007). Women as political participants: Psychosocial postconflict recovery in Peru. *Peace and Conflict: Journal of Peace Psychology*, *13*(3), 313–331.

Lederach, J. P. (2005). *The moral imagination: The art and soul of building peace*. Oxford University Press.

Meintjes, S., Pillay, A., & Turshen, M., (Eds.) (2001). *The aftermath: Women in post-conflict transformation*. Zed Books.

Nussbaum, M. C. (2000). *Women and human development: The capabilities approach*. Cambridge University Press.

Pasteur, K. (2011). *From vulnerability to resilience: A framework for analysis and action to build community resilience*. Practical Action Publishing.

PCS Project Counseling Service. (n.d.). *La mujer en el tiempo de la violencia política en el Peru: Testimonios y reflexiones* [Women in the time of political violence in Peru: Testimonies and reflections], (pp. 9–10). Consejería en proyectos (0285).
Pedersen, K. R. (2006). *Mobilizing poor people for active citizenship. United Nations Development Programme*. Oslo Governance Centre.
Petri, H. L. (1986). *Motivation: Theory and research*. Wadsworth Publishing Company.
Porter, E. J. (2007). *Peace building: Women in international perspective*. Routledge.
Reardon, B. A. (1993). *Women and peace: Feminist vision of global security*. University of New York Press.
Rigby, A. (2001). *Justice and reconciliation after violence*. Lynne Rienner Publishers.
Schaub, F. (2012). *The fear and anxiety solution: A breaking through process for healing and empowerment with your subconscious mind*. Sounds True.
Southwick, S. M., Bonanno, G. A., Masten, A. S., Panter-Brick, C., & Yehuda, R. (2014). Resilience definitions, theory, and challenges: interdisciplinary perspectives. *European journal of psychotraumatology*, 5, 10.3402/ejpt.v5.25338. https://doi.org/10.3402/ejpt.v5.25338
Suarez, E. B. (2011). *Surviving the "Sasachacuy Tiempu" (Difficult Times): The Resilience of Quechua Women in the Aftermath of the Peruvian Armed Conflict*. http://idl-bnc.idrc.ca/dspace/bitstream/10625/48930/1/IDL-48930.pdf
Sveaass, N., & Castillo, M. (2000). From war hero to cripple. *Peace & Conflict: Journal of Peace Psychology*, *6*(2), 113–133. https://doi.org/10.1207/S15327949PAC0602_2
The World Bank. (2001). *What is empowerment*. 2001/1: 39 (2002) Manual. http://web.worldbank.org/WBSITE/EXTERNAL/TOPICS/EXTPOVERTY/EXTEMPOWERMENT/0,contentMDK:20272299~menuPK:546167~pagePK:148956~piPK:216618~theSitePK:486411,00.html
Vella, D. (Ed.) (2001). In war, a women is often left alone: Women refugees in Eastern Africa". *War has changed our life, not our spirit: Experiences of forcibly displaced women* (pp. 47–70). Jesuit Refugee Service.
Young, I. M. (2011). *Justice and the politics of difference*. Princeton University Press.

Part IV

Conclusion

Identifying Threats to the *Vital Core*

7 Acknowledging the power gap

This chapter approaches the conclusion, including an analysis focusing on the meaning of power as an enclave within the word ***empower****ment*, and its implications in rural human development. Within identifying what was holding back rural women's empowerment in their agency from participation in decision-making, I found part of the answer is the power existing within the gap between the rural people and the system through their interactions within institutions. The point of departure concerned a power gap that explains the disconnection between those who make decisions at a systemic level and those affected by their decisions. The gap is filled with a dynamic interaction of objective/visible/concrete/cognitive and subjective/subtle/abstract/visceral communication forces between the system and the people. Those affected by decisions made at the institutional level in the capital of Lima were the rural people living in poverty, war, and searching for truth and justice. I recognised several power gaps that could have obstructed rural people's capabilities from having functionings to enable them access to decision-making in the aftermath. Power is not a static force in those who make the decisions, but rather, power is the abstract, subtle energy found in the gap between the subjects. This gap is a perception of power reinforced by subtle dynamics of experiences. Several identities were analysed within the chapters, and how their connotation in time and space either affected a disconnection or strengthened the relationship within the gap. Power is withheld within the gap, and transforming the dynamics of disconnections means new connections take place. Then, space enables the power to disperse, provoking powerful effects, namely *empowerment*. Figure 7.1 is a essential representation of perceiving these power gaps.

The power withheld on the top seems static, a wall filled with oppression, repression, and effort to overcome adversities in searching for truth and justice. It is so structured that it compresses and feels like matter. Empowerment could not occur when there was a lack of awareness of what permeates between the system and interactions within the rural population. The power gap could be reduced in its concreteness (density) and space (distance) if the system acknowledges its role in structuring through its institutionalism, affecting the people's capabilities, and rural women raise the consciousness of their agency when both (the system and people) interact with inner qualities of density and distance.

DOI: 10.4324/9781003450955-11

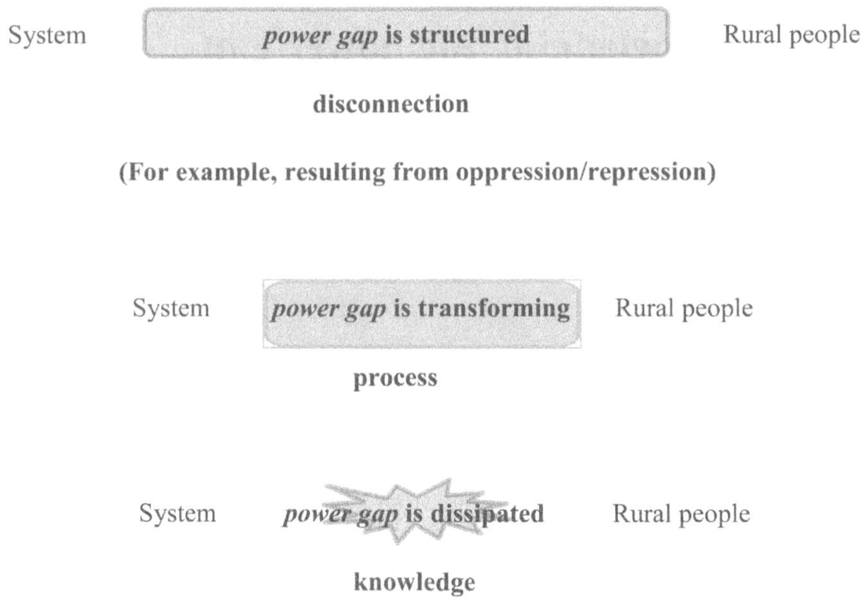

Figure 7.1 Disconnection, process, and knowledge. ©Rodriguez Carreon, V. (2014).

To analyse how this power gap is formed and transformed, this chapter opens a dialogue for awareness of the inner quality of the relations between the system and the rural people. Power is a dynamic relation of forces perceived as abstract or cognitive. Socially rooted, power as interactions exist in relationship. Foucault (1994) points out that power is in the action, individually or collectively, and "power is exercised only over free subjects". "Freedom must exist for power to be exerted" (Foucault, 1994, pp. 341–342). Therefore, power cannot be possessed, and it is decentralised in society (Dürmeier, 2012, p. 4). Power is also perceived as a structure of dichotomies; for example, marginalisation can obstruct and motivate the fight for justice. Over centuries, through the system's institutions, the peasant identity's connotation as powerless was first legitimised and internalised into the collective unconscious. Gaventa (2003) discusses the "duality structure" introduced by Giddens, where "people are free to act, but draw upon and replicate structures of power through their own actions". In this sense, institutions heavily rooted in systems with colonial structure become for the people the environment they breathe. The structures are also a dichotomy, the stimulus and the obstacle for people's agency in their empowerment. For example, how was oppression reproduced and mobilised through different dynamics of marginalisation, which pre-conditioned the effects of the peasants' agency during the conflict? At the same time, the injustice of marginalisation prevailing when dealing with conflict was felt as the catalyst for rural women pursuing justice.

The power gap's ramifications for the agency in transformation depend on the extent to which the capabilities are structured. Dürmeier argues about the capabilities concept of Amartya Sen:

> The transformation of characteristics into capabilities includes social, personal and environmental factors like physical handicaps, cultural values or political power relations. For Sen, political power is a transformation of capabilities, which changes the freedom of a person. An undemocratic government (for example,) can deliver food and nourishment by food stamps, but the personal freedom of self-production of foods or personal choice is prevented.
>
> (Dürmeier, 2012, p. 8)

I propose that capabilities for awareness work to reduce power gaps and enhance freedom for empowerment processes to transform human agency's stages. From Dürmeier's quote, I consider how the existent lack of awareness capabilities can obstruct the freedom of both the person and the system, where a political system under an undemocratic government negatively impacts people's agency by controlling the power gap. If people become alienated to fit their being and doings to those in decision-making for political power, then the development of their political agency would depend on the decision-maker's influence. Schneier confirms that many pressures imposed by society are attributed to those in power (Schneier, 2012, p. 195). The late Mendoza de Ascarza founder of ANFASEP was conscious of the powerful entities obstructing her search for justice, which, unless she became part of the decision-making process, any efforts to exercise her agency in her quest for justice would be a utopia. There were power gaps in the structural relationship between the perceived identities of rural people and the system. Power gaps were formed in the day-to-day interactions under oppression for centuries without *freedom from want*. The quality of agency then can be altered by disrupting consciousness, for example, when rural women seek the truth, displaying resilience in response to the confronted adversities. There is an opportunity in the liminal space when peasants are in the duality between the threat against their agency's *vital core* and their awareness of their capabilities to raise their consciousness.

This chapter oversees the power gap in structures of institutionalisation and explains the gaps that exist between institutions and the people. The *power gap* in institutions is fragmented into categories of *space* and *language* of authority. This fragmentation made by institutions forms a narrative of categories on how to be part of them by following their institutionalism beliefs, and when embodied, their concepts become institutionalised. Also, if the rule about the categories established by institutions interposes with people's values, the system sways over the relationship to control the power gap's distance to avoid people's resistance. Without consciousness, the power gap of institutionalism can change people's perceptions and provoke powerful effects to enable either empowerment or disempowerment dynamics.

When the power gap is in periods of transformation, human agency is also experiencing it by surrendering to the unknown and reversing the power over *uncertainty*. The gap is constituted by attachment to the known, to *certainty*. This power gap in relation to the unknown is wider when the agency is stuck in survival need. I focus on the agency transformation that can occur by reversing the negative attachments, physical or psychological.

The power gap is dissipated through the understanding of knowledge as the endeavour to achieve agency. The core of this gap lies in the concept of consciousness by experiencing *contemplation* and *introspection*. Both activities implicate the reflection necessary for the people who have been surrounded by traumatic environments, as those who experienced *llaki*. During contemplation the human *being* questions which *being* am I? And introspection is the capability of understanding how inner experience processes the *doing*: how am I *doing*? The consciousness of the *conscient agency* occurs when people are aware of the whole picture, and the people question their place in the world. It is a process of understanding their *vital core* functioning within their experiences. Acknowledging the inner and outer qualities of the power gap can enhance the freedom of the rural women's agency. Awareness in recognising the merging of *consciousness* and *capabilities* enhances the flourishing of the empowerment formation for a *conscient agency*.

The power of institutionalism

> We have no words to say what it is and how it feels to be workers, to be held in suspicion, to be ordered around by people who have more and who pretend to know more and who compel us to work according to rules they set and for purposes that are theirs, not ours. And we have no words to say all this because the ruling class has monopolized not only the power of decision-making and of material wealth; they have also monopolized culture and language.
>
> (André Gorz *in* Young, 2011, p. 192)

Those living under oppression experienced their cultures and languages being devoured. They were told they knew less, and what they knew was to be suspicious of the rule. Institutionalism in post-colonial countries is rooted in the exercise of control. Government institutions largely exercised and exerted dominium through ruling. Schneier (2012) explains how "throughout most of our history, people had no say in the laws that ruled them. Those who ruled did so by force and imposed laws by force" (Schneier, 2012, p. 195) on the people's lives. Institutionalism infrastructure withholds the authority of decisions to build a power gap by imposing the ways of *being* and *doings* on the people's needs and fears. During and after the armed conflict, the system used institutions to influence the physical, psychological, economic and familiar needs of victims-survivors.

The military discussed in Chapter 4—War and its rawness through cruelty and repression—operated under the narrative to attain the needs of people in relation to their security, and imposed rules for people to depend on the system's protection.

What is overlooked in these dynamics of interaction is the power gap mechanisms of being with doings, which under institutionalism the government ruled the needs of security to safeguard physical threats. This institutional power rests on people collectively conscious or unconsciously internalising functions; for example, Titus Stahl stated that "by giving people a reason to obey the legitimate demands of those who have the relevant powers", are accepted (Stahl, 2011, p. 351). For the system, categorising and monitoring people's functionings was part of controlling the reasoning power over the relationships' structure. There is a struggle among government institutions that want to relate differently with their citizens, yet they use the same colonial consciousness to build their people's affairs' architecture. Institutions categorise their experiences with their own interpretation and then perceive in terms of the institution's understanding. For example, rural people's capacities were not categorised by the active and resilient agents they were when they left their homes after the war. They became the displaced, disempowered victims monitored as vulnerable and portrayed as powerless. A narrative is constructed by conceptualising the new identities of those in displacement. When the peasants' identity had a disempowered connotation, relationships were created between the experiences of disconnection, forming power gaps of wants built on fear. Due to the system's lack of awareness of its own structure, where the reality of the people's affairs is not understood, identities with the oppressed outer image navigate institutionalism to prove their sense worthiness.

Krishnamurti said, "we are frightened also of the unknown–frightened of death, frightened of what lies beyond tomorrow. So, we are afraid of the known and afraid of the unknown" (Krishnamurti, 1969, p. 13). Here, Krishnamurti impels us to think that sometimes it is easier to keep control of this dual conflict by sticking to what we know and, at the same time, intrinsically there is a need to be open to the unknown. Rural women did fear the existing, what is known in their environment of scarcity and aggression, and through their experiences with the system have built knowledge in relation to what to anticipate from the institutions. When experiences start to be sensed as known and trustful, this uneasy confluence interaction with a system that oppressed collectives for centuries, bridging certainty and unity, is unknown and releases a sense of suspicion and, therefore, uncertainty—a suspicious, fragmented inner state that entered the collective biofield in oppression. The inner dynamics belong to the knowledge of the subjective, abstract, subtle, and sometimes invisible forces that require trust to be seen. In the case of oppression, when another similar negative experience repeats the pattern, the quality is embedded as an imprint of knowledge of the relationship with the system. This inner process obstructs the people's awareness about the system and its possible transformation and freedom for their agency to evolve.

With repression and aggression in conflict, rural women had feared what the military had done with the disappeared bodies and had feared when the night was approaching, and the *terrucos* were coming to abduct their sons or rape their daughters. They also fear the unknown, the process and what was involved when claiming justice at the institutions' offices and interacting with their representatives, including the fear of their death in the attempt to make the demands. Fear is not something concrete that is directly implanted by those institutions but is built up through policy measures such as the curfew during the State of Emergency, which prohibited reunions. Fear is the energy—sensing individual and collective—that unconsciously fills the gap between the people and the system. Consequently, the functioning of people's agency is constrained to exercise what they value and have a reason to value, as happened in the story of Rosario's mother, explained in the first chapter. The possibility of dying was real. The people living in the isolated areas of the highlands or the jungle got to know what happened to those who attempted to make claims, and the fear of what they knew, and the unknown, held them back in their freedom for their agency to evolve.

The power gap is formed in the daily interaction through the dynamic between institutions and people; for example, when people have been oppressed and repressed and feel they are not part of the environment, they see their dependence on the system where their "obligations" are followed without challenging them. Schenier argues that "Institutions include governments of all sizes, … [and] these institutions implement rules, laws, edicts – there are several terms – and sanctions for disobeying them and possibly incentives to obey them" (Schneier, 2012, p. 105). Consequently, the reason for obedience to institutions under the law is adapted from a norm. It gets its way into structuring institutional culture and, finally, becomes the "normal" for operating dynamics of oppression, repression, and aggression. Institutions fragment the affairs of the people, and in doing so, power gaps are constantly formed and reformed.

> Power must be understood in the first instance as the multiplicity of force relations immanent in the sphere in which they operate and which constitute their own organization; as the process which, through ceaseless struggles and confrontations, transforms, strengths, or reverses them; as the support which these force relations in one another, thus forming a chain or a system, or on the contrary, the disjunctions and contradictions which isolate them from one another; and lastly, as the strategies in which they take effect, whose general design or institutional crystallization is embodied in the state apparatus, in the formulation of the law, in various social hegemonies.
>
> (Foucault, 1978, p. 93)

I applied Foucault's view in the Peruvian case study to understand the interactions of people living in the Andes (*Sierra*) and Jungle (*Selva*) rural communities, with particular focus on the highlands, within the system's rules and obligations. The distance between the rural people and the system's rules embodies the gap. The interactions hold the "struggles" and "confrontations" that, as Foucault said, can evoke change. The gap withheld the power when the system categorised over structured experiences through compartmentalisation. This conscious and unconscious architecture becomes an institutionalism of habitual practices between the rural women and the system. For example, on security grounds, Quechua women requested protection from the forced recruitment they were enduring by the *terrucos*, and their needs were institutionalised through categories rather than including historical background of those who were "defending". The government implemented a policy known as State of Emergency, where the civilian population cannot count on the rule of law, as explained in Chapters 4 and 5. Several questions can be raised, such as how the government failed to safeguard civilians' human security. Who is the emergency policy designed for? Were there threats against the *vital core* of rural people protected? I wrote in Chapter 4 about the consequences of living under a State of Emergency where abuses were exercised in the name of security, something that happens in any conflict. There is a disconnection between the government representatives, the people they meant to defend, and the process that causes the power gap to be more distant, and dense. Capabilities for women's *conscient agency* in their empowerment formation also depend on a healthy society with government policies designed based on the human security approach, where the people become the decision-makers in their own destinies.

In Peru, power gaps were formed in contradictions. Different spaces, languages, customs, and cultures were between the rural people and those in charge of decisions within the system. By formulating laws and rules, institutions reinforce the legitimacy of those gaps, particularly when the people who aimed to be represented were not involved in the decision-making process to create the laws that safeguarded them. The challenge I see is what Varela et al. (1993) said: our representation of the world becomes what we think is knowledge of the world (p. 142). We need more consciousness to collectively inquire about our representation of the world. Our perception is the accumulated living experience that becomes consciousness and then the known of the world we see. Conscientising oneself confronts and transforms those embodied representations, feeling oppression and the perception of the encountered power gaps. Consciousness is needed from the oppression and repression exercised for centuries that broke the victims-survivors' trust in the Peruvian conflict's social, political, and economic institutions. Thomas Blass argues that "ordinary individuals could be induced by an authority figure to act destructively, even in the absence of physical coercion, and that it didn't take evil or aberrant individuals to carry out actions that were immoral and inhuman" (Blass, 2004, p. XVIII). The influence of authorities to which Blass refers is possible in a system where structures of oppression reproduce experiences through institutionalised mechanisms, building a multidimensional architecture without consciousness. At the same time, the "invisible agreements" on methods for social interaction were taken and expanded into other spheres. And under the umbrella of legitimacy, institutions with "immoral and inhuman" intentions adapt the delivery of policies to their convenience. An awareness of how power unfolds is required for people to acknowledge their agency position in decision-making.

I analysed in Chapter 3 how oppression is structured and, as argued by Young, it cannot be eliminated as a result of eliminating rules or imposing new rules. Oppression occurs when: "the systematic character of oppression implies that an oppressed group need not have a correlate oppressive group" (Young, 2011, p. 41). Oppression as an inner dynamic is like a poison that fills up the power gap with powerlessness where both the system and the people internalised. For example, inequality is a tool of oppression. It gives the Other a perception of hierarchy which fragments the concept of equality. The concept of inequality retains power when oppression is instructed by exercising identity differentiation. Differences that were categorised in hierarchical order allow power to be constructed unequally and in a fragmented way, and the identities that are at the bottom of the hierarchy are seen as powerless and those at the top are perceived as powerful. And powerless starts to be connoted negatively, and powerful positively, to the point that both sides internalise its rejection to the powerlessness within. A sense of powerlessness within oneself produces an abject inner experience, an internal Othering. Julia Kristeva's (1982) honest essay on abjection reflects that the abject confronts us, the object of original repression, inhabited by the sense of rejection to otherness (p. 134).

The power gap existing between the people and the decision-makers in hierarchical institutions is constructed in determined *space* and with a particular *language*. The point of departure from freedom from the power gap is in the awareness of how the power of space and language are withheld within the institution. Foucault writes that "we have to know the historical conditions which motivate our conceptualization. We need a historical awareness of our present circumstance[s]" (Foucault, 1982, p. 778). The argument focuses on the need to know how ways of *being* and *doings* in the present are conditioned by the past. It also urges us to raise consciousness to prevent blind spots in relation to the reproduction of institutional dynamics when becoming part of it, enhancing people's freedom from becoming institutionalised.

The *space* for decision-making

The *space* for decision-making holds power. Government institutions have different infrastructure bodies represented in a particular space in a determined location. In all the spaces decision-making is taken place. The space has inner qualities and influence, and it is influenced by perceptions of power from the people and from those working for the institution. The spaces are represented by physical infrastructure, symbols, logos, locations, and characteristics of those working there. If people in charge of the government institution recognise its features as their own, they risk becoming attached to the meaning of those characteristics. In those spaces, the experiences of *beings* are reproduced, reinforcing inner and outer qualities of their *doings* and others' *doings* until they become systemic. The continuous dynamic in the space creates more layers to function in certain ways according to the institution. While symbols, uniforms, and other visual representations form the persona, the subtle characteristics form a stronger bond. At the same time, the institution only exists because it represents and/or offers services to meet the needs of people, who at the same time need the organisation to meet their needs. In categorising people's needs, institutions' *language* creates a subtle distance and appropriates those needs; eventually, the decision-making on the needs no longer belongs to the people. The layers become more complex as hierarchical categories are created within roles and among other institutions. Power starts to intertwine like invisible cables through the abstract, the subtle and the inner Self across the social body's perception, where, with time, comes to a point where the people do not recognise themselves anymore in those categories.

For example, I have written in Chapter 4 about the cruel acts during conflict committed under the orders of those in authority. Then, in Chapter 5—In the search for truth and justice—I analysed how amnesty was given to those who were given orders under obedience to the authority and how they could avoid responsibility for the crimes committed. Milgam argues that "obedience is compelled by fear", and that "the essence of obedience consists in the fact that a person comes to view himself as the instrument for carrying out another person's wishes…the disappearance of a sense of responsibility is the most far-reaching consequence of submission to authority" (Milgram, 1974). In applying his argument, when institutionalised soldiers in the military executed immoral and inhuman acts under obedience, they became the author of the wound, distancing their sense of responsibility for both the wound and for the connection to humanness. There is a fragmented relation with spaces used for training and/or indoctrination where misanthropy internalises to obey in committing barbaric acts, as analysed in Chapter 4. The absence of inner qualities and skills is seen in the disconnection in *space* (infrastructure) and *language* (culture) of the power gaps. The lack of empathy is generated in metamorphoses when that authority perceives itself as the owner of the *space* and *language*. There are, however, other qualities for decision-making; for example, the neuroscientist Francisco Varela said:

> this trait [empathy] of our lived body plays a decisive role in the manner in which I apprehend the other, not as a thing but as another subjectivity similar to mine as alter ego. It is through his/her body that I am linked to the other, first as an organism similar to mine, but also perceived as an embodied presence, site and means of an experimental field. This double dimension of the body (organic/lived)…is part and parcel of empathy, …the means to social conscious life.
>
> (Varela, 1999, p. 81)

Therefore, if under authority, committing harms is decision-making reasoned as a duty, and not crimes, traits like empathy that connect to human values and morality are wounded. Overwhelmed by the capacity to sense oneself as someone who values and has a reason to value, when perpetrating hazards, the perpetrator *vital core* is disconnected. Mark Johnson argues:

> We are beings whose identities emerge and develop in an ongoing process of interactions with our physical, interpersonal and cultural environments. To function successfully within these changing environments our reason must be expansive, exploratory, and flexible. The locus of our moral understanding is thus our imaginative rationality (a *human*, rather than *Universal*, Reason) that allows us to envision and to test out in imagination various possible solutions to morally problematic situations.
>
> (Johnson, 1993, p. 219)

Sensing oneself moral is fundamental to feeling connected to our sense of Self. Frame (2015) writes about the unseen wounds of soldiers injured in the age of barbarism. They are not visible yet impact the soul. To see the invisible requires a consciousness cultural approach, including spiritual. In Peru, when institutions encourage the repression of moral injuries, as explained in Chapter 4 in the space of training with hazing, humiliation and shame, for example, the trainee's capacities to connect with the abstract, the subtle and the subjective are disconnected as well. Then, the power gap becomes more dense, and more distant. In contrast, I found in rural women's rationality in understanding the process, for which recognition of moral choices should be praised. For example, Lima, the capital of Peru, is the main geographical *space* for decision-making with government affairs. It is where power is held. The power gap between the oral encyclopaedic speakers and the government was noticed by a rural woman when she needed to claim for justice to search for answers about her family member's disappearance. The late Mendoza de Ascarza, learnt in her truth and justice seeking process the need to travel to Lima, to the *Palacio de Justicia* (Palace of Justice), an unknown institutional space for her at that moment. It was unknown as infrastructure, as well as the interactions with its customs and culture. She had to learn about the institutional processes and people's functioning in that *space* and how they dealt with *language*. Angelica met there with the authority in charge of the institution, in his space. With every visit to the institution, her knowledge on the dynamics grew. Her inner agency contributed the power gap, with the system becoming less dense for her and less distant. The power gap with the Palace of Justice gradually reduced its perception of strength in power over the Self. The institution, at the same time, did not learn anything about the peasants. The government, unaware of how historically the State consciousness was affected, did not see to reduce the gap. They feared the unknown, and instead of embracing it, they used abuse of power to increase it and maintain control over the *space,* and the *language*. Freire said that people are "no longer satisfied to watch, they want to participate … At first, the elite react spontaneously. Later, perceiving more clearly the threat involved in the awakening of popular consciousness, they organize … [Because] they must preserve at all cost the social 'order' in which they are dominant. They cannot permit any basic changes which would affect their control over decision-making" (Freire, 1973, pp. 13–14). These *spaces* in institutions can enhance or hold back the empowerment formation in the agency of the rural people. Building on Freire's argument, in the Peruvian experience, new rules were created to keep rural women from isolated communities at the bottom of the identity hierarchy. Nevertheless, the late founder of ANFASEP developed new capacities on how to deal with the system which were explained in Chapters 5 and 6. Raising consciousness helped her to reduce her perception of the power gap, and not see it as concrete wall, but as a system's failure to see the agency of the peasants.

The *language* barrier

Language is more than an idiom, it is the way we think, learn, and communicate. Described by Karlberg as "the ways we think and talk about a subject influence and reflect the ways we act in relation to that subject" (Karlberg, 2005, p. 1). *Language* is "arguably one of the most powerful tools in world politics today" (Hudson, 2010, p. 1). For example, I perceived using *language* to hold power when I attended one of the public hearings' seminars. Representatives of many rural communities and urban organisations affected by the armed conflict met in the salon of the Congress Palace. The meeting was for victims-survivors' representatives to express their experiences in relation to reparations. The parties interacting in the seminar were the victims-survivors' representatives from different regions, NGOs that were there to support the victims-survivors and the High Level Multisectoral Commission (committee known as CMAN) representing the State. As explained in Chapter 5, the understanding of reparations priorities for the government differed from the community. Victims-survivors wanted to communicate their frustrations about reparations, and the committee wanted to communicate what they had done for victims-survivors. The *language* became a power gap. Here, I am not referring to Quechua, the main language of victims-survivors, or Spanish, for the commission's representatives. It seemed there were two different ways of talking, acting, and reacting. The *space* in which this interaction occurred was the authority's *space*; the CMAN and the Congress Palace were government entities. The NGOs were the bridge between victims-survivors and the CMAN, and the CMAN at the same time was the bridge between the State and the victims-survivors. The experiences of interactions among them relied on the perception of each group, where they either extended the bridge or formed and reformed a breach.

As I analysed the power in different structures where oppression has dominated for centuries, the difficulty can be seen in assessing the chances for rural women to enter the power gap of consciousness, when they were connoted as the powerless with unknown parameters and conditions for decision-making. Women's agency is powerful and unrecognised in the power narrative. The power discourse has been attributed to the action and the one in charge of the action, to be visible. While, on the contrary, agency unless exerted is unknown. Rural human development has developed survival, passive, active and conscient agency, yet this is poorly acknowledged in the literature. It needs a new *language* where the experiences of oppression and powerlessness that rural people's agency embodied are seen. The historical wound needs to be seen. However, powerlessness is something that nobody wants; it is negatively connoted and associated with inaction as if not *doing* something was existential. Therefore, *being* is seen as not acting and wanting to be invisibilised, including its *language*, as if it belongs to the powerless.

Is it possible to reverse the power gap of uncertainty?

Institutions build a top-down approach through processes that exclude people in the decisions and become attached to the reason behind those impositions. Oppression over time creates "uncertainty" through *exclusion* and *attachment* in both the system and the people. The feeling of emptiness from uncertainty about circumstances and not knowing aspects of a situation, as was the case for rural women during the conflict, threatens the *vital core*. Living in poverty and war brings uncertainty, as seen in Chapters 4 and 5; and sensing uncertainty (freedom from the known) is at risk of being wounded, which wounds are manifested through control (power over) or submission (powerlessness). The continuous threats against their peace of mind capabilities kept their inner struggles to sense themselves alive. The struggle started to hold the power, yet it was a power formed in disconnection with their *vital core*. In a struggle to sense oneself, what is needed is awareness of the struggle. Thus, the apprehension of uncertainty clashes with the nature of experiencing peace of mind and the sense of Self.

How, then, is peace manifested in times of uncertainty? How can the transition from *active* to *conscient agency* be developed? Victims-survivors want to know the locations of the bodies of those who disappeared, what happened to them, and how they can claim justice from the perpetrators. Attaining complete knowledge about what happened would create for the people a sense of certainty and security. The other extreme is the complete absence of knowing, not having any control in the process and, therefore, not being able to overcome the feeling of being trapped in sensing uncertainty. Over the years of oppression and repression, the control over the power of uncertainty has been given to the system. Structured oppression wounded the freedom of the known in rural communities. On the other hand, the rural women who experienced *llaki* under repression need consciousness to restore their connection with their *vital core*. In a way, what is impelled here is to consider identifying the threat against the *vital core*, then people can own their own reality, rather than given the power of that internal process to the system. In the following paragraph, Juddi Krishnamurti explains:

> If I were foolish enough to give you a system and if you were foolish enough to follow it, you would merely be copying, imitating, conforming, accepting, and when you do that you have set up in yourself the authority of another and hence there is conflict between you and that authority. You feel you must do such and such a thing because you have been told to do it and yet you are incapable of doing it. You have your own particular inclinations, tendencies and pressures which conflict with the system you think you ought to follow and therefore there is a contradiction.
>
> (Krishnamurti, 1969, p. 17)

Krishnamurti said that contradiction exists between the people and their system's tendencies. However, rural women were asking for the truth from the system because the State had responsibility to answer for those who disappeared. It is paradoxical, the disempowered connotation of ways of being a peasant, a Quechua, since colonial time perpetuated in different ways until the time of war. Chapter 3 narrated how oppression existed for the peasants through marginalisation, which persisted before the conflict. Rural people were unnoticed. The absence of State institutions become visible when the Truth and Reconciliation Commission discovered that 70% of those who died did not have a formal identification. During conflict, repression added to oppression, and aggression did not ignore the peasants. In contrast, rural people from the highlands and the jungle were the largest percentage of those who died, their lives were not safeguarded. It seems unavoidable to be conditioned to conform and to accept a system as Krishnamurti describes, in particular when an identity is associated negatively with powerlessness. Its negative connotation becomes part of the subjects' lives, and progressively, the person becomes the reflection of the given category of marginalisation in which they live. Do the people become dependent on the system, or did the system became dependent on having people oppressed? The enquiry here is how governments that did not embody and embed the capabilities to safeguard their rural population can, in times of war, identify the threat to their people's inner condition.

When institutions fragmented the whole of the people's experiences created more distance with their people. Their hierarchical system of authorities controlled the *space* of *language* of the experience. The power gap became more plausible in war, as uncertainty grew for the people and the system. For example, when perpetrators of violence are uncomfortable with uncertainty, made the power gap denser by intensifying their cruelty. Also, when rural women were uncertain about what they will eat the next day (lack of *freedom from want*), and uncertain about what can happen when the *terrucos* attack their communities (lack of *freedom from fear*).

Is the desire to be at the decision-making table the freedom to be certain? Or is it becoming conscious of the influence of the power gap that gradually brings freedom from being certain? Freedom gives space for "peace of mind" to enable a connection with the sense of Self. Maturana and Varela (1987) discussed that people always look for certainty. They said, "we tend to live in a world of certainty … this is our daily situation, our cultural condition, our common way of being human" (Maturana & Varela, 1987, p. 18). I am not saying that the desire for certainty in cases of disappearance should not be acquired. In contrast, consciousness of the dynamics for the search for truth and justice—explained in Chapter 5—is necessary to shift the perception of systemic repression perpetuated against the victims-survivors. The capabilities of certainty depend on the system discharging its duty of responsibility, but also in acknowledging the intrinsic process of agency. The step to invert the power in this inner conflict between the duality of certainty and uncertainty—that rural woman confronted when they were affected by the *llaki*—lies in witnessing the effects of that power in the development of their own agency. For example, the decision-makers of the system are aware when a situation causes fear to the institution, this information gives the system a sense of uncertainty. There is power in witnessing the fear and free the wound by sensing uncertainty. Hellmuller (2013) confirms that people within such a vulnerable context are not passive recipients, because they take and interpret on their own what they perceived (p. 223).

Understanding uncertainty lies in the awareness of two main points: the recognition of inclusion and the need for detachment from conditioned negative attachments. For the rural women, core beliefs are a process of thinking about the experiences attached to their disempowered connoted identities and traumatic events that felt as definers of their identity, as explained in Chapters 3 and 4. Exclusion and inclusion are withheld in the power gap and once the system and the people are part of that power through witnessing, exclusion is disintegrated, and inclusion is dissipated. The power gap holds the "binary oppositions" that Foucault refers to among others as the "drives/consciousness" (Foucault, 1978). While the analysis is in a different context, focusing on the potential inner and survival threats to the agency is relevant. For example, women's *active agency* in survival environments discussed in Chapters 4 and 5. When consciousness is raised with inclusive capabilities, the system will understand more of the inner development in *conscient agency*, and be able to offer a human security environment.

Ethics for inclusion: recognising the power gap of exclusion

Concentration creates exclusion when it becomes the centre. For example, if we concentrate our thoughts on one particular experience, we exclude the rest of our experiences, and then we are unable to see the whole panorama of experiences. We had centred one experience, and there were others outside the centre, on the margins, and then concentration became the centre, which confused the panorama. In Krishnamurti's (1969) philosophy, if we apply this analysis of concentration and exclusion to violence, not only outwardly in war but also the confusion in ourselves inwardly. Then, we can learn to observe the confusion but with consciousness. We include all experiences to see the whole picture.

There is a need to recognise that we are ethical and reflective human beings; and therefore, we should enhance inclusion. Then, the power gap would not have the capacity to oppress and repress as explained in Chapters 3 and 4. For the Peruvian case study, the power of inclusion means to consider the power of all human agency, and their contribution to structuring an inclusive system. A system that includes the safety of the people's *vital core* recognises human security comes from those who own their reality for decision-making. The effectiveness of this power depends on the level of including the affairs of their citizens. Amartya Sen argues that:

> As competent human beings, we cannot shirk the task of judging how things are and what needs to be done. As reflective creatures, we have the ability to contemplate the lives of others. Our sense of responsibility need not relate only to the afflictions that our own behaviour may have caused (though that can be very important as well), but also relate more generally to the miseries that we see around us and that lie within our power to help remedy.
>
> (Sen, 1999, p. 283)

There is a need to understand the power in inclusion as a sense of having moral responsibility to reduce the power gap. Morality as an action, for example, can reduce the gap in the power of misery. Sen quotes Rawls on moral powers: it is "a capacity for a sense of justice and for a conception of the good", in fact, the role of values is extensive in human behaviour. He adds, "it is the power of reason that allows us to consider our obligations and ideals as well as our interests and advantages. To deny this freedom of thought would amount to a severe constraint on the reach of our rationality" (Sen, 1999, p. 272). Sen here makes a call to the sense of duty of the State and its system to recognise the ethics of inclusion. A nation's system is composed of its citizens and the institutions of its government (Sen, 1999, p. 274). There is a relationship of power between the State and the social structures. But, from the international relations theory, Wight et al. (1978) discuss that power seeks security without referring to justice, and refers to vital interests regardless of common interest (p. 293). Therefore, power distribution is not considered an option because people need security, which is seen as vital to survive.

One way of reducing the uncertainty by inclusion can be if the State provides capabilities for including the rural agency in decision-making in relation to the policies on dealing with the disappeared. Rural women hold the experience and therefore the knowledge of the meaning of disappearance which was explored in Chapters 4 and 5. Instead of the State fragmenting exclusion as part of poverty and disappearances as part of the conflict, exclusion needs to be intertwined in one simultaneous observation. Multidimensional dynamics can build a State that recognises inclusion and empower the knowledge capabilities of the system. Knowledge, as previously argued, reduced the uncertainty of the people and the State. When uncertainty affects the matters that people and State value and have a reason to value, it also compromises the *vital core* of the people or the *vital need* of the State. Alkire argues that "elements of the vital core are fundamental human rights which all persons and institutions are obliged to respect or provide" (Alkire, 2003, p. 3).

The dichotomy of attachment: power gap in the transition of vulnerability

Poverty and conflict had irreparable consequences and the people affected by the oppression and repression experienced weakness, yet at times they also experienced strength. "This dual aspect of emotions – … [has] destructive and non-destructive sides" (Dalai Lama, 2013. p. 119). "War is only a brutal desecration of their memory … loss and losing … grief, failure, brokenness, numbness, uncertainty, fear, the death of feeling, the death of dreaming. The absolute, relentless, endless, habitual unfairness of the world" (Roy, 2002). What does loss mean to victims-survivors? In a patriarchal culture, being vulnerable is perceived as a weakness. However, it also can be the strength to keep fighting and exert *passive agency*, as explained in Chapter 6 (Participation for peace). How did rural women perform through their different roles as mothers and political negotiators impact their communities and, consequently, their system? It is important to highlight that when in chaos, rural women were active and reactive to unsafe circumstances. They became more vulnerable to *llaki*, for example, when they were attached only to memories of traumatic experiences. I analysed how attachment to a particular memory (Krishnamurti, 1972), can compromise the *vital core*. Therefore, awareness is necessary to enhance this transformation in acknowledging rural women's agency in vulnerable times.

The strength within the vulnerability power gap is found when people's courageous acts and experiences transform and weave their inner and outer agency. It is in that vulnerability that strength was withheld, for example as mentioned by the known *mamá Angélica*, it was in the unfairness of pain she had seen that she encountered the motivation to keep going (A. Mendoza de Ascarza, personal communication, 2009). She experienced the pain of others, empathy for it, and provided her compassion and resilience. A therapeutic process can enhance healing through suffering for those who are aware of their uncertainty. As Pasteur argues, the link between vulnerability and resilience is finding ways to understand uncertainty through positive strategies and decisions that bring well-being to livelihoods (Pasteur, 2011).

The power of the people: when the awareness of the experience becomes knowledge

The experience becomes knowledge (Freire, 1973; Krishnamurti, 1969; Varela et al., 1993). And such knowledge, according to Freire, is conditioned, in the case of peasants (Freire, 1973, p. 108). I applied this argument to the Peruvian case, and I found that rural people experiences were also conditioned by the years in which they accumulated knowledge. Once they add more awareness into the process of how their acquired experience became their knowledge, their rural agency will also have more capabilities for their decision-making. The power of perception is a shift in the conceptualisation of the experience of those who own the meaning.

I argue that human agency is held back by the lack of awareness of their *consciousness-capabilities*. If *consciousness-capabilities* are embodied and embedded in the people and the system, their internalised awareness will guide their freedom. For example, in Chapter 3, I analysed how the system structures an environment of oppression, which the rural people embodied through marginalisation and exclusion. They become embedded in the system of oppression when the internalised marginalisation affects their *vital core*. Then, the symptoms of oppression gradually become visible, but without consciousness, they are reproduced subconsciously, creating a pre-conditioned context to sense the threat of marginalisation. At this point, I inquire, can we free our agency from the language of marginalisation situated in the collective subconscious? We can raise our awareness of the inner qualities of our *situatedness*. For instance, if people's capabilities and system awareness are enhanced, that consciousness will enable *situatedness* for *conscient agency* to be free. *Situatedness* is *conditio sine qua non* (Lindblom & Ziemke, n.d.), condition without which it could not be. Survival is the ultimate *conditio sine qua non* for all needs (Webel & Galtung, 2007, p. 11). Freedom, then, from *survival agency*, is the essential condition for our *situatedness*. At the same time, *situatedness* can be understood as an awareness of people's conditioned perception of knowledge of their system, which lives in the collective subconscious. "Knowledge is made always in a relational context" (Maturana & Varela, 1987 p. 174) through communication, and for humans through language. At the same time, language is *conditio sine qua non* for the experience (Maturana & Varela, 1987 p. 231).

This understanding of the relationship of how people relate to language to acquire knowledge is essential to understanding oral agency and its expression through technological artifacts. While people and system can be traced back to Aristotle in his analysis of ethics and politics. The aim is not to focus on the analysis of theory and the relationship of concepts; rather, it is to understand the concepts in relation to the people's experiences and the system, in order to understand the process of empowerment of those whose experiences come from a difficult background. Agency within the power gap is ***empowered*** by people when the knowledge of consciousness is not only known intellectually but exercised to raise a more conscientious agency through reflection on introspection and contemplation. Through these reflecting practices of awareness, the oral agency in rural human development can be understood as knowledge connected to their Environment, where both are not separated. The next section aims to understand the people's knowledge and its capabilities to enable the consciousness of rural human development to flourish with freedom.

Contemplation

Contemplation is a state of consciousness. In contemplation, it is the capability to consciously create an inner space to process detachment from the conditioned context that was led by the perception built on lived experiences with the Self, the Other, and the Environment. In Chapter 1, The point of departure, I explained how Freire (1985) suggested admiring each word in the theme. I applied his argument to contemplate the words that became part of the theme, noticing the meaning withheld within their history before they become concepts and before they interact with the theme. Extending our contemplating capabilities shift the perception of distance in the power gap, by seen the gap between the concepts and experience closer. For example, in exercising contemplation, I notice what concepts encapsulate what lives in me, and I sense the relationship of each meaning in the wholeness of the theme, knowing my *situatedness*, in the *here* and the *now*. This exercise of disengagement to engage "properly" is an experience which is sought in order to achieve clarity. As argued by Freire:

> Human beings are active beings, capable of reflection on themselves and on the activity in which they are engaged. They are able to detach themselves from the world in order to find place in it and with it. Only people are capable of this act of "separation" in order to find their place in the world and enter in a critical way into their own reality. "To enter into" reality means to look at it objectively, and apprehend it as one's field of action and reflection. It means to penetrate it more and more lucidly in order to discover the true interrelations between the facts observed.
>
> (Freire, 1973, p. 105)

Freire argues later that in the peasants' case, this exercise in understanding reality makes them realise that they are part of the reality rather than seeing themselves as the transformers of that reality (Freire, 1973). In Chapter 6, women peasants in Peru transformed their motivations to find the truth once they were in contemplation of their suffering. Let's recall the case of the founder of ANFASEP, when she mentioned that to see the dead bodies where the corpses of the people were dumped (*infiernillo*) and contemplate the unfairness of it, all gave her more strength to seek for justice. Contemplation requieres awareness, therefore, her inner experience of contemplating unfairness have been resourced from her resilience and ignated a shift in her agency. In the same chapter, resilience was analysed as a capacity for coping with pain in which agency is manifested through participation in survival agency. Rural women also contemplated their suffering as a collective. For inner support, they accessed their resilience capability built through their ancestral way of being and doings in introspection with the Environment. As Suarez, who also analysed the rural Quechua women in poverty and conflict, said: "The resilience of Quechua women appears to be sequential and specific to the *events*, *processes* or *invariant conditions* they confront. I am deeply convinced that Quechua women are not powerless; they cope with their experiences of trauma and loss in unique ways through their increased political engagements, they have begun to challenge gender stereotypes" (Suarez, 2011, p. 195). The analysis of Suarez about Quechua women is framed under Galtung's three concepts of violence: direct violence (events), structural violence (processes), and cultural violence (invariant) (see more, Galtung, 1990; Suarez, 2011).

Resilience, however, is challenged when they experience *llaki*, as explained in Chapter 4, for which further inquiry is needed as what is the language of contenmplation in rural human development. The process of contemplating, then, can bring women a greater awareness of the conditions that put them in the place that they are in. Contemplation to raise awareness of ourselves contributes to a sense of clarity of what role we play in our environment, and bring consciousness to safeguard the *vital core*.

Introspection

Introspection is the reflection necessary to know oneself and enhance the capabilities of consciousness to be free. Maturana and Varela (1987) have done an extensive analysis on knowledge in which they argued that "reflection is a process of knowing how we know. It is an act of turning back upon ourselves. It is the only chance we have to discover our blindness and to recognize that the certainties and knowledge of others are, respectively, as overwhelming and tenuous as our own" (Maturana & Varela, 1987, p. 24). The understanding of the introspection process brings a conscious cognitive approach. In the psychology area this has been evaluated as the positive motivation. How an individual can function fully is what Carl Rogers or Abraham Maslow called the motivation for growth and self-actualisation. The inner process of motivations and how we reason the beings and doings of what we value and have a reason to value was analysed in Chapter 2, *Consciousness-Capabilities*. Also, a "motivation is fuelled by the reasons why it's good for you to change" (Schaub, 2012, p. 57). The experience of learning through doing the action, may not be understood or accounted for, but in leaving the mind open, the process through reflection produces conscious knowledge. I explained in Chapter 6 how learning was uncomfortable and painful for rural women when their agency was transforming and evolving. I noticed in observing this process that the pain was beyond cognition and included the embodiment of what they were internalising.

Introspection helps to understand the knowing of the inner world. It is a process of observing how actions become determined circumstances (Maturana & Varela, 1987, p. 174). This knowledge of the self cannot always bring happiness in cases where negative actions produced from the past come to the surface—an extreme example of transgression might be for Shining Path members or military personnel who committed barbaric acts. However, awareness brings an understanding of the relation among facts and how the actions within those facts were reproduced. It is that awareness that brings certainty to the uncertain, a clarity to the reality, to understand why actions were committed so measures can be taken to not reproduce those transgressions. As Maturana and Varela explain (1987), "what we intend … is to be aware of what is implied in this unbroken coincidence of our being, our doing, and our knowing. We shall put aside our daily tendency to treat our experience with the seal of certainty, as though it reflected an absolute world" (p. 25). This reflection will help those in despair to come to terms with their traumatic experiences, for example, when mothers living in poverty and war with precarious conditions had to escape to the mountains because they feared for their lives; or any example given as the consequences of living with the lack of freedom from want and freedom from fear explained in Chapters 3 and 4.

This process of internal analysis can be applied to the system to enhance the need for the recognition of agency in rural women's courage for survival in the Peruvian conflict. The State have a duty to focus on their people. Amartya Sen argues the importance in the understanding of this agency:

> Understanding the agency role is thus central to recognizing people as responsible persons: not only are we well or ill, but also we act or refuse to act, and can choose to act one way rather than another. And thus we–women *and* men–must take responsibility for doing things or not doing them. It makes a difference, and we have to take note of that difference. This elementary acknowledgement, though simple enough in principle, can be exacting in its implications, both for social analysis and for practical reason and action.
>
> (Sen, 1999, p. 190)

I accept the limitations of going deeper into the essence of introspection function to shift inner agency stages. However, I want to point out that this process of introspection can change the State *situatedness* within a system. Introspection will involve understanding the stages of agency, a process that will enable the State, as Sen said, to make an elementary recognition of people as agents. Ultimately, the State can offer a change in its system by acknowledging the disempowered dynamics in the *language* and *space* of its institutions.

Conclusion: awareness of the *being* and *doings* in em*power*ment

Figure 7.2 illustrates what is explained in this chapter in relation to the awareness of the power gaps which hold back rural women's *conscient agency* for their empowerment formation.

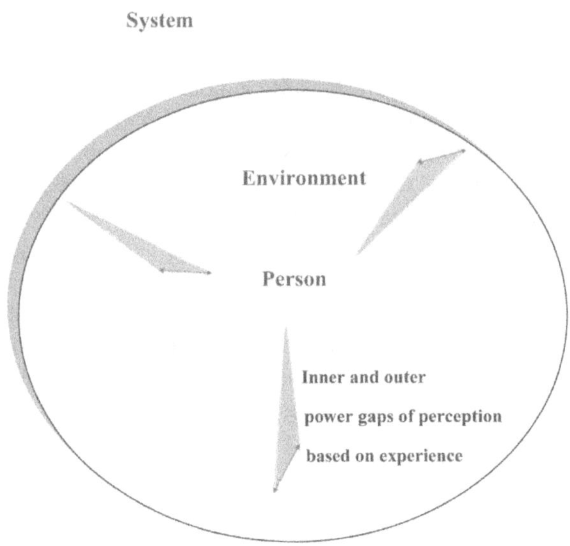

Figure 7.2 Awareness of inner and outer power gaps. ©Rodriguez Carreon, V. (2014).

How were these gaps formed, assuming that power lies within the gap? The gaps were built by experiences and the re-enacting of those experiences between the *person* and the *system* and vice versa—for example, the experiences of oppression, repression and injustice explained in Chapters 3, 4, and 5. I found that when the gap is minimal, the empowerment capabilities are higher, and those involved have more chance of dissipating that power through their awareness of the existing disconnections. The gap can be small, large, or even intermittent. During conflict, the gap could change depending on the everyday experiences. The distance can be reduced by the system's awareness of its influence on the power gaps. Less awareness of the power gap represents less freedom; therefore, their freedom for rural human development lies in gradually exercising consciousness of their experience of human agency.

When the person and the system connect genuinaly, power is dissipated; because in that moment, there is not power, just awareness. Krishnamurti said, in the moment power appears, control appears and control breeds violence (Krishnamurti, 1969). It is in that violence that torture in conflict is exercised. In the gap of physical power, exerted force is possible through fragmentation. Therefore, the power delivered in torture is "overpowering"; the person is exercising "power over" another person (s), and yet the person is labelled as "powerful". For the person who has been tortured, the experience is "powerlessness", but the person is labelled as "powerless". The narrative and the connotation of power are problematic. While victims of torture are powerless, why is it that being "powerless" receives a negative connotation? As analysed in Chapter 4, *terrucos*, and soldiers, were trained to be extremely aggressive and portrayed as courageous in narratives, even if they used extreme violence. Therefore, the *terrucos* should not be given the category as someone "powerful" because someone who cannot recognise internally his or her own actions is not exercising power. A shift is necessary in the narrative of power. Power is when awareness of the power gap shift its empowerment formation.

This chapter highlighted the importance of the acknowledgement of the power gap and how the people and the system can benefit from it. If understanding the interaction of power dynamics operating in *space* and *language* applied by institutions is manifested in relation to the rural women's agency and the system, raising consciousness can provide common knowledge to transform agency into a *conscient agency*. Awareness of who we are and what we do, as well as our *being* and our *doings*, would not come from the imposition of the system through their institutions; it would come from the knowledge of the experience. Particularly in oppression, Paulo Freire explains how to raise knowledge.

> Knowledge is not *extended* from those who consider that they know to those who consider that they do not know. Knowledge is built up in the relations between human beings and the world, relations of transformations, and perfects itself in the critical problematization of these relations. In order to discuss any kind of technical question with peasants, they must see this question as a "distinct perception." If it is not this, it must become this. Whether it is a "distinct perception" or not, the peasants still must in both cases apprehend the interplay of relations between the "distinct perception" and other dimensions of reality. The effort required is not one of the extension but of *conscientização*. If it is successfully carried out, it allows individuals to assume critically the position they have in relation to the rest of the world. The critical taking up of this position brings them to assume the true role incumbent on them as people. This is the role of being Subjects in the transformation of the world, which humanizes them.
>
> (Freire, 1973, pp. 109–110)

Concientização is a Portuguese word that can best be translated as raising consciousness. Freire argues that this process of *concientização* for the peasants is necessary for them to become the owners of their own history. In order for this process of *concientização* to be acquired, the "power over" exercised in State security, must shift to a comprehensive human security approach. Alkire confirms, "human security is 'people-centred'; it focuses the attention of institutions on human individuals and their communities" (Alkire, 2003, p. 8). When the citizens are neglected or abused their connection with power is fragmented. The lack of inclusion of rural people in decision-making processess is reinforced by not recognising their contribution to the State's structure. Thus, if people are not included and are unaware of the system's structure, the State system will have a fragile foundation which brings human insecurity. In that case, "the outward social structure is the result of the inward psychological structure of our human relationships, for the individual is the result of the total experience, knowledge and the conduct of man. Each one of us is the storehouse of all the past" (Krishnamurti, 1969, p. 13). The person and the system need freedom from the experience of a conditioned social structure.

Consciousness-capabilities can be understood as the process of "understanding of the totality of life in which every fragmentation has ceased … that very understanding of the structure of thinking is its own discipline which is meditation" (Krishnamurti, 1969, p. 115). The meditation process can permit the consolidation of the wholeness lost in fragmented and conceptualised experiences. Knowledge is conceptualised to understand the experience. For example, in Chapter 2, I analysed how conceptualisation is rooted in the experience: poverty would not exist if experiences of living in poverty did not exist, conflict would not exist if experiences of living in conflict did not exist, transitional justice would not exist if experiences of justice in transition from war (for the Peruvian case) did not exist, and so on. However, fragmentation occurs in the conceptualisation process, and its split creates conflict. Once people are attached to fragmentation, depending on the level, they are not able to see clearly the wholeness of their agency's experience. For example, government institutions conceptualise a person's experiences under different affairs. In the aftermath of conflict, different organisational bodies analysed the experiences of the rural Quechua people, yet their experiences were filtered through areas of the discipline. This method is problematic because it is only part of the truth, and ignores the parts of the wholeness in experience influenced by other parts that might be under different disciplines. This is the conflict of perceptions in which power gaps erupt. In order to disseminate the power, the wholeness needs, as Juddi Krishnamurti proposed, the discipline of meditation which "demands an astonishingly alert mind … that kind of discipline is fluid, free, not the discipline of conformity" (Krishnamurti, 1969, pp. 115–117). Meditation, or the practice of consciousness, enables the *space* and *language* to connect with oneself's agency capabilities.

References

Alkire, S. (2003). *Conceptual framework for human security* (Working Paper 2). Centre for Research on Inequality, Human Security and Ehtnicity, CRISE. University of Oxford. https://assets.publishing.service.gov.uk/media/57a08cf740f0b652dd001694/wp2.pdf

Blass, T. (2004). *The man who shocked the world: The life and legacy of Stanley Milgram.* Basic Books.

Dalai Lama (2013). *Beyond religion: Ethics for a whole world.* Rider books.

Dürmeier, T. (2012). *The analysis of power with the theories of Amartya Sen and Pierre Bourdieu.* Hamburg University. http://Quechuan.assoeconomiepolitique.org/political-economy-outlook-for-capitalism/wp-content/uploads/2012/06/Duermeier-2012-Power-Capabilities-04.pdf

Foucault, M. (1978). *The history of sexuality. Volume I, an introduction* (R. Hurley, Trans.). Penguin Books.

Foucault, M. (1982). The subject and power. *Critical Inquiry*, *8*(4), 777–795. http://www.jstor.org/stable/1343197

Foucault, M. (1994). *Power: Essential works of Foucault 1954-1984* (R. Hurley et al., trans.). In J. Faubion (Ed.). The New Press.

Frame, T. R. (Thomas, R.) (Ed.). (2015). *Moral injury: Unseen wounds in an age of barbarism*. UNSW Press.

Freire, P. (1973). *Education for critical consciousness*. Sheed and Ward.

Freire, P. (1985). *The politics of education: Culture, power and liberation*. Bergin and Garven.

Galtung, J. (1990). Cultural Violence. *Journal of Peace Research*, *27*(3), 291–305. http://www.jstor.org/stable/423472

Gaventa, J. (2003). *Power after Lukes: An overview of theories of power and their application to development*. First draft. http://Quechuan.powercube.net/wp-content/uploads/2009/11/power_after_lukes.pdf

Hellmuller, S. (2013). The power of perceptions: Localizing international peacebuilding approaches. *International Peacekeeping*, *20*(2), 219–232. https://doi.org/10.1080/13533312.2013.791570

Hudson, N. F. (2010). *Gender, human security and the UN: Security language as a political framework*. Routledge.

Johnson, M. (1993). *Moral imagination: Implications of cognitive science for ethics*. The University of Chicago Press.

Karlberg, M. (2005). The power of discourse and the discourse of power: Pursuing peace through discourse intervention. *International Journal of Peace Studies*, *10*(1), 1–25. http://www.jstor.org/stable/41852070

Krishnamurti, J. (1969). *Freedom from the known*. Harper One.

Krishnamurti, J. (1972). *The impossible question*. Krishnamurti Foundation.

Kristeva, J. (1982). *Powers of Horror: An Essay on Abjection*. Columbia University Press.

Lindblom & Ziemke (n.d.). *Social situatedness: Vygotsky and beyond*. University of Skövde. http://cogprints.org/2517/1/Lindblom.pdf

Maturana, R. H., & Varela, J. F. (1987). *The tree of knowledge: The biological roots of human understanding*. Shambhala.

Milgram, S. (1974). *Obedience to authority: An experimental view*. Tavistock.

Pasteur, K. (2011). *From vulnerability to resilience: A framework for analysis and action to build community resilience*. Practical Action Publishing.

Rodriguez Carreon, V. (2014). *Empowerment formation: women's agency for participation in decision making within the poverty and conflict context, case of rural Peru* [Unpublished Doctoral thesis]. The University of Sydney.

Roy, A. (2002, October 1). "Not again". *The Guardian*. http://theguardian.com/world/2002/sep/30/usa.iraq

Schaub, F. (2012). *The fear and anxiety solution: A breaking through process for healing and empowerment with your subconscious mind*. Sounds True.

Schneier, B. (2012). *Liars and outliers: Enabling the trust that society need to thrive*. Wiley and Sons.

Sen, A. (1999). *Development as freedom*. Oxford University Press.

Stahl, T. (2011). Institutional power, collective acceptance, and recognition. In H. Ikaheimo & A. Laitinen (Eds.), *Recognition and social ontology* (pp. 349–372). BRILL.

Suarez, E. B. (2011). *Surviving the "Sasachacuy Tiempu "[difficult times]: The resilience of Quechua women in the aftermath of the Peruvian armed conflict* [Doctoral thesis, University of Toronto]. https://tspace.library.utoronto.ca/bitstream/1807/31951/1/Suarez_Eliana_B_201111_PhD_thesis.pdf

Varela, F. (1999). Steps to a science of interbeing: Unfolding the dharma implicit in modern cognitive science. In G. Claxton, G. Watson, & S. Batchelor (Eds.), *The psychology of awakening: Buddhism, science and our day-to-day lives* (pp. 71–89). Rider.

Varela, F., Thompson, E., & Roch, E. (1993). *The embodied mind. Cognitive science and human experience*. MIT Press.

Webel, C., & Galtung, J. (2007). *Handbook of peace and conflict studies*. Routledge.

Wight, M., Bull, H., Holbraad, C., & Spence, J. (Eds.). (1978). *Power politics*. Bloomsbury Academic.

Young, I. M. (2011). *Justice and the politics of difference*. Princeton University Press.

8 A hologram perception

In the statement by Paulo Freire:

> The point of departure of the movement lies in men [people] themselves. But since men [people] do not exist apart from the world, apart from reality, the movement must begin with the men [people]-world relationship. Accordingly, the point of departure must always be with men [people] in the "here and now", which constitutes the situation within which they are submerged, from which they emerge, and in which they intervene. Only by starting from this situation—which determines their perception of it—can they begin to move. To do this authentically they must perceive their state not as fated and unalterable, but merely as limiting–and therefore challenging.
>
> (Freire, 1972, p. 57)

Freire suggests a person is part of her/his reality and that the person only exists in relationship to it. The point of departure for the person is embodied in their *being* and *doings*, arising from the interaction with their reality, which is the "here and now". The quality of interactions between the inner and outer self conditions the person's understanding of his/her environment. If the person does not abide by the system because he/she senses injustice from the institutions, then the person enters into a relationship with the system to bring change. The person or group then, through a shared goal, act based on what he/she/they value and have a reason to value. Applying Freire's understanding as part of the conclusion brings further light to how I explored stories like Rosario at the point of departure for this book. The freedom to attain knowledge for transformation may lie in the practice of the "here and now" by Freire, and in the meaning of meditation by Krishnamurti. To be free to experience empowerment and exert a *conscient agency*, it is necessary to be in the present moment, aware of the interactions with the environment/system in my participating body, mind and spirit.

I have found it challenging to work out how to conclude in a way that will be fair to each experience of the rural women living with triple adversity. How can their historical marginalisation, deprivation, exclusion, and discrimination experienced since colonialism free the condition to access their freedom from want (poverty) and freedom from fear (war)? To make a fair account, I present the conclusion as

DOI: 10.4324/9781003450955-12

a hologram because I have found it to be the only way to see wholeness. The other way to see the totality of life that Krishnamurti refers to requires the reader to apply inner sensing practices and embodiment to connect from a different source that is not only intellectual. A hologram perception sees the parts in each dimension simultaneously; there is no linear order. What Senge et al. (2004) called "presence", "understanding how parts and wholes are interrelated" (p. 5). Yet, it is not feasible for me to do a hologram to sense the consciousness of rural people on paper. It will require what John Paul Lederach (2005) called a moral imagination, "the capacity to imagine something rooted in the challenges of the real world yet capable of giving birth to that which does not yet exist" (p. IX). Figure 8.1 represents the multidimensional being and doings within the main experiences (resilience, participation, seeking the truth, and knowledge) exerted by rural women in relation to their system (oppression, repression, injustice, and lack of acknowledgement).

I explored the different theories that became the pillars for understanding the analysis. The framework explained in Chapter 2—*Consciousness-capabilities* under the umbrella of human development and human security—supports the study of functionings between the experiences of rural women's agency and the system. In Chapter 3—Poverty and its forms of oppression—and in Chapter 4—War and its rawness through cruelty and repression—I explored living in poverty and war under the umbrella of human development and human security narratives rather than a traditional development and security discourse to focus on a human-centred approach. I considered under which circumstances *capabilities* and *consciousness* are more likely to occur or not occur if threats or barriers are identified in war. This diagram does not show a conclusive interweaving leading to a solution, but it is intended to highlight the obstacles constantly forming, evolving, and transforming the agency in the empowerment process of rural people. A hologram perception in which can be seen the wholeness of the agency's experience enables the interpretation of the disconnections that contribute, and to what extent, why rural women's empowerment was withheld.

In the graphic, the environment surrounding rural people comprises primarily living in poverty and war. As discussed in Chapter 3, oppression meant experiencing marginalisation for ethnicity (language and culture), gender, and class; through exclusion across all dimensions of social, educational, political and cultural. Then, the war brought them human insecurity, repressive measures with intense forms of aggression. The shock for the Truth and Reconciliation Commission was to acknowledge that more than 70% of rural Peruvians who died during the war had never registered an ID. The interconnection among the concepts of the book is represented in Figure 8.1.

Living in chronic human insecurity leads to the exhaustion of people's sense of being. Described in Chapter 4, armed conflict represented the rawness of violence through cruelty when the Shining Path and the military repression terrorised the rural population mentally and physically. The terrorists and special commandos of the military went through dehumanising methods of discipline, repressing the body's instinctual capabilities to prepare for war. Soldiers went through repressive forced assimilation experiencing pain, hazing, rape, and other grotesque experiences that

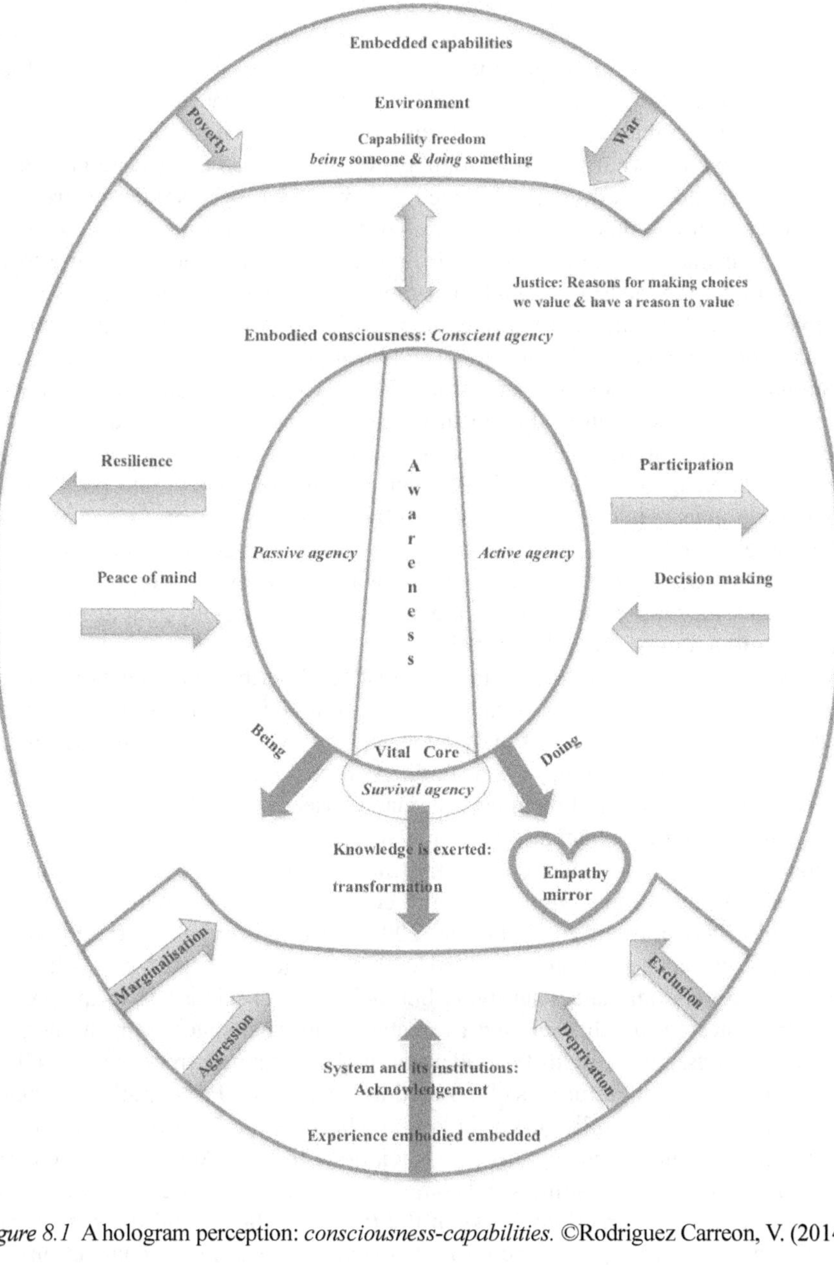

Figure 8.1 A hologram perception: *consciousness-capabilities.* ©Rodriguez Carreon, V. (2014).

led to humiliation and the suppression of shame to themselves. These inflicted actions, sensations, and feelings in the name of discipline were exerted under legitimate measures, for example, the State of Emergency. Without the rule of law, they could break into houses in the middle of the night, exert the repressive anger accumulated into the various forms during training, and be protected legally. The *terrucos* also underwent represive trainings. Under coerced and forced indoctrination, many children and youth abducted and recruited to be soldiers were given weapons to inflict pain on themselves by killing and committing abhorrent crimes against their human condition and their family members, where they were forced to show loyalty to the Shining Path. This dehumanisation happens in any war, and intensifies the disconnection with people's *vital core*. Without freedom from fear, rural people were on *survival agency*. The two arrows in the graphic at the bottom, marginalisation and exclusion, are strong frames for deprivation and sustainers of oppression. Rooted in institutionalism created power gaps before, during and in the aftermath of conflict. The rural population left their communities, exerting their inner experiences of resilience (built when in *passive agency*) and participation (exercising their *active agency*), as shown by the two arrows on each side. At the same time, their feelings were moving inwardly, with in-power-moments seeking truth and justice, yet they were actively in their agency for decision-making. Inner movements for their peace of mind are seen as *passive agency*, but agency is not exerted without movement that goes inward.

Chapter 5, In the Search for Truth and Justice, describes the transitional justice process in Peru. As well as the Truth and Reconciliation Commission report investigating the past experiences of the victims and survivors of the conflict (I referred to them as victims-survivors). Their findings, rural, peasants and/Indigenous people are those who suffered the most inhuman consequences of the war to the point they doubted their own human condition. Part of sensing the capacity of humanness died in the exposure to the cruelty of any war. In Peru, the people of the highlands, who were excluded and marginalised since colonialism, were also the most affected during the 20-year armed conflict (1980-2000). Victims-survivors' capacity to cope with resisting death is often associated with resilience. However, capabilities for resilience are not only about surviving but including consciousness of the process. Women protected the children and overcame many challenges from the negative connotations associated with their identity, which stemmed from cultural alienation to the dominant culture, as elaborated in Chapter 6—Participation for peace. They participated in family survival plans and became the decision-makers in their communities. Rural women formed associations to re-create spaces to interact, make decisions for their families and communities, and seek those who disappeared.

The increased interactions between the rural population and the system enabled women to participate in new roles, revealing the different connections and disconnections called "power gaps", explored in Chapter 7—Acknowledging the power gap. Power in the system is "labelled" on the collective unconscious as overpowering, power over, powerful, powerlessness, and powerless. It is essential to review the role of institutions, where power gaps are constantly forming in *space* through *language*.

Raising consciousness for empowerment in human development is about human flourishing, but it is also witnessing the inhuman experiences that formed *survival agency*. Bringing awareness is the opportunity to reconnect with vulnerability by witnessing the experiences of oppression, repression, and aggression. By gradually strengthening the letting go of the need to be certain, in Krishnamurti's (1969) words, is freeing the Self from the known. Raising awareness to understand transformation through perceiving the totality of life is not something that happens overnight. In rural human agency in Peru, consciousness is necessary to witness *llaki*, a Quechua word, to express idioms of distress. *Llaki* was a response that visibilised their *survival agency* and kept them alive in their transformational process. Wars affect individuals but also are collective experiences that impact all areas of life. Therefore, inner transformation in empowerment processes can also be collective. Rural women in Peru, owning their own transformation, they are part of this world.

Part of the conclusion to the findings on what held back rural people's agency in participation in decision-making, is in the *fractus*, the Latin word for being fragmented and broken—which conditions the mind, body, and spirit. A hologram perception shows how *fractal* experiences in war create and recreate partial perceptions simultaneously, and these perceptions ignite motivations or compromise the sense of empowerment. For Capra (1997) parts are patterns of an "inseparable web of relationships" (p. 37). Conditions for agency transformation can be created in each part by the people themselves, yet the power is in not giving the inner possibility of empowerment transformation to the system, instead taking responsibility for it. However, when the *vital core* has been affected by oppression and/or repression, awareness is not possible until it is freed to enable the capability of connection with its agency. I mentioned the different experiences that pushed the *vital core* of human beings to the point of being absent from their own humanity. Therefore, the responsibility is systemic. The systems play a role in the re-organisation of humanity. Institutions, through space and language, are the enablers of conditions for the empowerment of people. In this process, I found four main stages of agency, which were explained and applied in the analysis throughout the chapters.

Passive agency in poverty and its forms of oppression combines coping with the environment (resilience), assimilating new skills (capacities), and accumulating core beliefs in their lived-experiences. Rural women adapted ways of being to become embedded in their system in order to survive. In poverty, they built capacities adapted their agency to their vital core needs to survive conditions of disempowerment encountered when their identity was marginalised. In passive agency, movement is inwardly and yet unseen to the outer.

Survival agency in war and its rawness through cruelty and repression was explained with examples of the grotesque violence that happened in the rural areas of Peru. War dynamics threatened the coping mechanisms adapted to their passive agency when experiencing oppression. In survival, there is no well-being when the vital core is fragmented from the Self. Human agency is compromised, disconnected from the core beliefs formed by what people value and have a reason to value. In war, the capabilities to exercise agency are exacerbated, disturbing people's ability to exert their capacities and threatening their inner capacity to sense themselves, in other words, disembodied. Confused perception is a symptom of a jeopardised pattern for survival. Bohm (2002) explains fragmentary perception as an "unconscious habit of confusion" of what is seen and manifested in a subtle way (p. 22). As Bohm said, the confusion is subtle. Therefore, its survival embodiment can go unnoticed by the culture of the outer world when its *vital core* is compromised from the ability to sense the Self.

Active agency in the search for truth and justice, taking action in systems of oppression and repression becomes a matter of survival. In this stage, the inner agency is acknowledged to a new level of consciousness, and the learnings of survival are exerted, transforming inwardly from survival passive to survival active. Rural people's adaptation to survive metamorphosed through oppression and repression constantly challenges their human condition. In this stage, rural women were triggered by the memories of traumatic experiences. At the same time, some women's *vital core* were not compromised in their agency with *llaki*, and they could use their capacities to participate in activities that impacted their lives and their communities. At the same time, it is essential to acknowledge that this stage of coping in survival is not sustainable. Resilience evolved by being active and participating, where the inner and outer self may or may not be connected. In *active agency*, people might suppress what they were feeling to be able to act, therefore, it is active, but not necessarily conscious.

Conscient agency in participation for peace happens when an *active agency's* impact process is conscious. In war, considering the inner struggles rural people experienced while participating in their search for peace, there is a level of collective consciousness raised when they acknowledged their capacities and weaknesses as the driving forces supporting their actions. This agency stage evolves when there is awareness on the level of recognising oneself's capabilities for formal decision-making. The power gaps visibilised the quality of interactions between the people and the systems' institutions. In Peru, the rural people, Indigenous and/or peasants' way of exercising knowledge through their oral culture rooted in the Earth's cosmovision, expressed through texture and other artifacts technologies, were suppressed over centuries, denying its existence in consciousness by the power over. They constantly struggled during the conflict, and their active agency continued to be denied being seen by their system, which intensified and deepened the invisibility in layers of the collective unconscious through repression. The stagnation state in their agency needs the awareness of the experience to be free, for rural Quechua women to feel intrinsically secure in their *vital core*, empowered and flourishing in their agency. To be conscient requires contemplation and introspection, as seen in the previous chapter. I want to highlight that in moments of empowerment, awareness also emerged for rural people amid violence, from their ancestral inseparable relation with their Environment.

Each stage of agency introduced has forces working towards and against the cores that either motivate or obstruct rural people's agency. In this case, finding those who disappeared will give women a sense of fairness in their search for the truth. The environment is the context and, like the system itself, is a structural force. Most of the time, individuals think they do not have power within these forces. Indeed, institutionalism seemed to have a life of its own and was consciously and unconsciously overpowering through space and language. The analysis was made as the chapters unfolded of how much these forces influenced the empowerment formation, either contributing or holding back rural agency.

Each time the system's forces influenced the possibilities for peasants' agency, the capabilities for awareness also shifted, for both the person and the system. While living in poverty and war, rural women's agencies evolved from needing to know how to participate to entering formal decision-making in their communities. For example, ANFASEP founder late Mendoza de Ascarza's long journey to find answers to her son's disappearance evolved in her agency when consciousness was raised. Then, rural agency capabilities enabled people's motivation to raise consciousness. Consciousness-capabilities is a process that is not stagnating but transforming and reinforcing the human agency dynamics.

Rural women's agency in the process of justice in transition is acknowledged through the storytelling of their participation and decisions made in their search for the truth. Uncertainty affected the relationship with the Self. Then, the level to which the human spirit in their *vital core* is affected will depend on the power gap between the person's inner and outer sense of awareness. An inclusive system is necessary to bring awareness to the relationship between forces that obstruct the acknowledgement of people's participation in formal decision-making. Therefore, systems must work in rural human development and human security to identify the threat to their system, which is the population's *vital core*. When I refer to the development of the human as the centre of society, I am considering, as Schaub said, the wholeness of the human being "consist[s] of body, mind, and spirit" (Schaub, 2012, p. 5). For capacities to evolve in a human-centred approach, people need the freedom to be and to do what they value, have a reason to value as agents of change, and transform their own environment by participating in the decision-making that impacts them. The system and its institutions (which reproduce other systems simultaneously) need a hologram perception process to integrate rural people's agency through participation and decision-making and transform their country's common well-being.

As part of the conclusion, I recommend the merged concepts of *consciousness-capabilities* to enable empowerment through the agency awareness to do and to be free to make moral choices for reasons that we and our systems value and have a reason to value, in which awareness of the lived experience is free of a conditioned perception. It requires:

> The capacity to imagine ourselves in a web of relationships that includes our enemies; the ability to sustain a paradoxical curiosity that embraces complexity without reliance on dualistic polarity; the fundamental belief in and pursuit of the creative act; and the acceptance of the inherent risk of stepping into the mystery of the unknown that lies beyond the far too familiar landscape of violence.
>
> (Lederach, 2005, p. 5).

Let's consider the web of relationships in which the person and system are in tune with what they ought to transform. For this to happen, the person and the system need to step into the mystery of the unknown that Lederach refers to. This step will free the space to sync the person/system agency with the capabilities to raise consciousness. When applying *consciousness* and *capabilities* in rural human development, barriers can be recognised, and strengths can be acknowledged. Acknowledgement is key for empowerment to be enhanced. For people who went through the inhuman conditions of living with feelings of guilt of someone who has been abused to the point of rejecting their own identity, healing is necessary—yet healing in the rural human development cosmovision. For our socio-cultural collective healing, it is essential to integrate the knowledge from oral agency, as well as their idioms' ways of expression in *consciousness* and *capabilities*. The new system needs to include oral culture's communication on inner development and human flourishing, such as their idioms of emotions. Emotions are very powerful, as noticed by the Dalai Lama; they come and occupy in full our consciousness (2013, p. 131). The awareness of those emotions is essential to the "living in the present" (Varela, 1999). The system needs to improve its institution's human development by looking deeply at how they categorised the lived experiences of the rural people living in geographically isolated areas, and how they influenced the negative connotation of people identities.

This work concludes that trauma inflicted through oppression, repression, and aggression needs collective awareness. This research identifies that the threat of war is beyond the number of deaths. The capability and consciousness to sustain life require an integrative awareness of each current system's fragmentation. To collectively address the aftermath of an armed conflict, I resonate with this question "if trauma yearns to be voiced yet is unspeakable, from what point of view or perspective will the narrative of that story emerge?" (Hübl & Shridhare, 2022, p. 10). The approach in response to discerning the collective narrative is systemic, which, as Capra (1997) said, means to be contextual (p. 37), and dynamic. We need a symphony awareness in which language is not static; it is an expression dancing with the whole, in Bohm's words, a "rheomode", an undivided movement where we can attempt to see the world through one aspect but only as a non-local abstraction of the whole (Bohm, 2002). I conclude by acknowledging that the consciousness of agency requires systems to be in tune with their capability of holding the splits within, of what Scharmer (2018, p. 134) called, the divides: spiritual/cultural, social, and ecological—or also, as mentioned in previous chapters, a disconnection between Self/Self, Self/Other(s), and Self/Environment.

The body is the place of the senses (Varela et al., 1993, p. 65). Therefore, shifts required to sustain peace must minimise decision-making rooted in cognitive dominant approaches. The practice of being attuned is embodiment. It enables a genuine generative ability to respond in decision-making to what is experienced in context, in the here, and the now.

To be in the present is something that deepened my research. I found that the inner quality of attunement in oral agency was revealed by the Andean rural people as they grazed their animals on a hillside and exerted a particular quality of their relationship between their agency and their Environment. Expanding awareness "filled with knowledge of a very different quality than cognitive understanding", I recognised during my action research journey with the peasants as a spiritual resource sustaining the rural people's resilience (more in Green, 2023). In Hübl's (2023) words, attunement, "like listening, is essential to authentic relating". It's a mature relationship. Attunement means listening and perceiving another person with all of our senses engaged, which includes regulating the speed of how we perceive someone to whom we are relating at that particular moment. "It's about developing the capacity to align your frequency to their frequency, your perception to their perception, your biofield to their biofield" (Hübl, 2023, p. 46) I found learning about collective attunement is critical, particularly in countries with rural communities that share a common history of poverty and war. The "collective psyche is holographic" said Thomas Hübl (2020, p. xxiii) in his book, "Healing collective trauma, a process for integrating our intergenerational and cultural wounds". His teachings strengthen the need to apply new lenses when evaluating the aftermath of protracted conflicts, where we see the person and the system, the individual and the collective as a whole.

What is needed for social embodiment in systems when they are in periods of transition? Lederach's (2005) work suggest the quality of space and relational capacities that, through time, can sustain processes (p. 97). It needs a system where their institutions' space and language acknowledge their rural population, and the system develops their embodying relating capabilities with consciousness. More specifically, it suggests the need for a collective witnessing of the systems exercising power over entire communities' lived experiences of marginalisation and exclusion and the way institutions embodied negative connotations of rural people's gender, ethnicity, and deprivation of economic development (class is understood as resources with power over economic networks, a hierarchical perception introduced with colonialism). When citizens sense their political system has betrayed them, being able to trust again requires an expression that comes in a new form of embodiment in the social field. Overcoming the suspicion that separates them from their political needs requires a system sensing the "language" of what has been fragmented. Like many original oral languages, Quechua is not linear (Rodriguez Carreon, 2023). The cultural cosmovision is attuned to a more extensive body of expression, which includes the land and other living organisms as their environment. Trust is a movement unspeakable with words, and yet, it communicates in relationship. Enabling it is a craft that involves the inner and outer agency where trust is not a command but an expression, a sound motion landing reciprocated as a licence to believe.

The development of agency in empowerment formation and transformation proposed in this book needs the presence of *consciousness* and *capabilities*. A hologram perception based on *consciousness-capabilities* enables the awareness of our lived-experience to be free in order to be someone and to do things with freedom. It is an approach focused on identifying the threats in systems that holds back the empowerment of agency in rural women living in poverty, war, and the aftermath. This work invites us to continue the exploration of inner and outer agency and its transcending expression of peace. Consciousness in rural human development starts with acknowledging communities' oral agency connection with their Environment, visibilising their expression of its human technology. Through sensing the rhythm quality of awareness when in communication with the Self, the Other, and/or the Environment, human agency evolves. Applying *consciousness-capabilities* lenses to free the inner and outer development of human agency is to safeguard the vital core of rural people's human security as *conscient agents*.

References

Bohm, D. (2002). *Wholeness and the implicate order*. Routledge.

Capra, F. (1997). *The web of life: A new synthesis of mind and matter*. Flamingo.

Dalai Lama (2013). *Beyond religion: Ethics for a whole world*. Rider books.

Freire, P. (1972). *The pedagogy of the oppressed*. (M. Bergman Ramos, Trans.). Penguin Books.

Green, M. (2023). Listening to the Pause in Peru. In *AlterNatives: Matthew Green on How Trauma – Personal, Collective, Ancestral – Reverberates Through Communities, Strengthening Their Powers*. https://www.thealternative.org.uk/dailyalternative/2023/4/2/matthew-green-alter-native-trauma-peru

Hübl, T. (2020). *Healing collective trauma: A process for integrating our intergenerational & cultural wounds*. Sounds True.

Hübl, T. (2023). *Attuned: Practicing interdependence to heal our trauma—and our world*. Sounds True.

Hübl, T., & Shridhare, L. (2022). The 'Tender Narrator' who sees beyond time: A framework for trauma integration and healing. *Journal of Awareness-Based Systems Change*, *2*(2), 9–27. https://doi.org/10.47061/jasc.v2i2.4937

Krishnamurti, J. (1969). *Freedom from the known*. Harper One.

Lederach, J. P. (2005). *The moral imagination: The art and soul of building peace*. Oxford University Press.

Rodriguez Carreon, V. (2023). Liminality: Change starts within, *Challenges*, *14*, 25, https://doi.org/10.3390/challe14020025

Rodriguez Carreon, V. (2014). *Empowerment formation: women's agency for participation in decision making within the poverty and conflict context, case of rural Peru* [Unpublished Doctoral thesis]. The University of Sydney.

Schaub, F. (2012). *The fear and anxiety solution: A breaking through process for healing and empowerment with your subconscious mind*. Sounds True.

Senge, P., Scharmer, O., Jaworski, J., & Flowers, B. S. (2004). *Presence: Exploring profound change in people, organizations, and society*. SOL Crown Business.

Scharmer, C. O. (2018). *The essentials of Theory U: core principles and applications*. Berrett-Koehler Publishers, Inc.

Varela, F., Thompson, E., & Roch, E. (1993). *The embodied mind: Cognitive science and human experience*. MIT Press.

Varela, F. (1999). Steps to a science of interbeing: Unfolding the dharma implicit in modern cognitive science. In G. Claxton, G. Watson, & S. Batchelor (Eds.), *The psychology of awakening: Buddhism, science and our day-to-day lives* (pp. 71–89). Rider.

Index

For Product Safety Concerns and Information please contact our EU representative GPSR@taylorandfrancis.com
Taylor & Francis Verlag GmbH, Kaufingerstraße 24, 80331 München, Germany

www.ingramcontent.com/pod-product-compliance
Lightning Source LLC
LaVergne TN
LVHW021127300926
849160LV00010B/899

9781032586533